INTERVENTION IN HUMAN SERVICES

INTERVENTION IN HUMAN SERVICES

EVELINE D. SCHULMAN

Professor, Community Mental Health,
Morgan State College; Consultant in Training Paraprofessionals;
Program Administrator, Evaluation Design Community Programs
for Developmentally Disabled,
Maryland Mental Retardation Administration;
formerly, Chairperson, Department of Psychology;
Coordinator, Mental Health Technology,
Community College of Baltimore,
Baltimore, Maryland

THE C. V. MOSBY COMPANY

Saint Louis   1974

Library of Congress Cataloging in Publication Data

Schulman, Eveline D
    Intervention in human services.

    Bibliography: p.
    1. Social service.   2. Interviewing.
3. Counseling. I. Title. [DNLM: 1. Community
mental health services. 2. Social service,
Psychiatric. WM30 S386d 1974]
HV40.S393   361   73-19740
ISBN 0-8016-4370-8

Cover photo by Jerry Sloan,
City of Chicago, Department of Human Resources.

VH/M/M   9   8   7   6   5   4

TO MY HUSBAND

without whom this book
could not have been written

PREFACE

The idea for this text originated with my students. After bombarding students with books and pamphlets since the initiation of the Mental Health Technology program at the Community College of Baltimore in 1967, the inadequacy of material pertinent to human services became frustrating. Our program was planned so that training in knowledge would be coupled with training in skills in the classroom laboratory and in the associated field experiences. Many texts supplied the didactic aspect; some satisfied the experiential aspect for skill training. However, no books integrated the didactic with the experiential and could be used with undergraduates from a wide variety of backgrounds.

Handouts were the starting point from which this book was written. Students used these handouts as their informational and skill-training guides in conjunction with books about human behavior and about the human services field. The content was revised on the basis of students' criticisms and will continue to be revised, hopefully, with suggestions from the expanded number of readers using this book.

The subject matter gradually took form as students used the handouts as part of their laboratory training. Students' reactions to the materials as they practiced their skills in their field experiences, and later when the first Mental Health Technology class graduated in 1969, provided ideas for needed content and skills.

The general style of this book is conversational and words associated with professional jargon are defined. A glossary pulls together all of the words defined within the text, plus a few others that may need additional explanation. The reading level is focused on a heterogeneous population ranging from individuals offered short-term training within an agency to those in four-year programs related to human services. Students at a two-year community college, the Community College of Baltimore; students in a four-year college, Morgan State College in Maryland; and one group of students trained in a short-term plan at an Inner City Community Mental Health Center have tried out most of the exercises and have read most of the contents.

Knowledge and skills are integrated in this book. Explanations of concepts precede exercises for developing the skills associated with the concepts. A layered approach to training begins with the simpler, basic skills of *O*bservation, *R*ecording, and *R*eporting, continues into *I*nterviewing, and then to the more complex skills based on interviewing, namely, *C*ounseling—the ORRIC approach. The organization of the book allows for beginning at any one of the layers of the acronym ORRIC, depending on the trainee's existing knowledge and skills. Thus the more advanced student may omit some of the beginning, Part One, which focuses on observation, recording, and reporting and, perhaps after a brief review, go right into interviewing, Part Two or Part Three. Part Four, counseling, builds on the other three parts of the book, and yet can stand alone to provide refining of counseling skills for individuals who are already working in counseling situations.

During the writing of this book, the thought has consistently popped up, "Is this something you have created?" It is particularly difficult to acknowledge all of one's resources and references when footnotes are omitted and a special effort is made to insert few bracketed references so that the students' flow of reading will not be interrupted.

Therefore I apologize for any oversights of unrecognized borrowed thoughts.

Space limitation permits words of appreciation to only a few of the students who have waded through the rough drafts of the manuscript. Three students are particularly worthy of mention for their worthwhile suggestions: Yvonne Ilgenfritz, Connie Reinwald, and Benjamin Smith. I am also indebted to Dr. Harold McPheeters and to James King of the Southern Regional Education Board, from whom I gained a great deal of understanding and also the opportunity to present some of the content of this book at the regional and national training meetings they sponsored. Others who have directly or indirectly helped with their comments and encouragement include Dr. John True and Dr. Carl Young from the Center for Human Services Research, Johns Hopkins; Dr. Ralph Simon, National Institute of Mental Health; Dr. Paul Shoffeit, Antioch College, Maryland; and the many directors of two-year human service programs who have urged me on to complete this book.

Some special people offered encouragement, understanding, and material assistance in completing the manuscript. My son Ken offered valuable suggestions and assisted in compiling the glossary and references. My son Mark provided a sounding board, which I needed, and discovered the origins of some of the ideas expressed. Kary, my daughter-in-law, patiently and imaginatively created the illustrations. There are no words to adequately express my appreciation to my husband Sol for his encouragement and manuscript-long assistance.

For Lawrence Powell, a special thank-you for conscientious help in preparing and putting the pages of the manuscript in order. Thanks are also due to Lou Anne Hamel, who typed the manuscript.

Eveline D. Schulman

CONTENTS

PART ONE

OBSERVATION: THE BASIS FOR HUMAN SERVICES

LOOK AND LISTEN FOR THE WAY TO GO

CHAPTER ONE
HUMAN SERVICES— THE CENTIPEDE

The many segments of services to humans. (Photo by Jerry Sloan, City of Chicago, Department of Human Resources.)

1

Mike T., M.D., put his stethoscope to Matthew's chest and listened to the irregular heartbeat. Dr. T. thought to himself, "This fellow needs further examination . . . EKG [electrocardiogram] and. . . ." Dr. T. is a human service worker.

2

Paul waved to the group of young men and women standing or sprawling on the walk in front of the drugstore. One of the young men called out, "Hey, Paul, commere. . . . " Paul is a former "delinquent" and high school drop-out trained in an 18-month intensive course to be a "youth worker." Paul is a human service worker.

3

Ms. M. has her master's degree in clinical psychology. She is employed by the department of education to work with children at an inner-city elementary school. Poor attendance, low achievement, and "behavior problems" are some of the school's concerns. Three students in a mental health technology program at a local community college are doing their field work (practicum) at this school with Ms. M. as their supervisor. Ms. M. is a human service worker and the three students are in the process of becoming human service workers.

4

Anne repeated her request and added, "Look here, I'm a social worker, Ted's a psychologist, Bud's a psychiatric nurse, Helga's a mental health associate, Larry's a health assistant—and all others, different titles . . . but . . . listen . . . we're all doing similar things here. What are we arguing about? Let's talk about how we can serve the people's needs more effectively . . . no matter what our degrees may be." This community health center is staffed with human service workers.

• • •

What is a human service worker? Answering questions with more questions is irritating, perhaps. However, the questions "How is the phrase 'human services' explained?" and "How is a professional self established?"

do open the way to consideration of present-day ideas about the helping approach as well as who may become a helper.

HOW IS THE PHRASE "HUMAN SERVICES" EXPLAINED?
Philosophy, procedure, and definition of human services

The concept of human services is a philosophy that stresses care for the *whole* individual and his relation to his environment. It poses a straightforward notion—an individual is influenced by what happens around him; the same individual influences what happens around him. These two kinds of influences make the individual what he is at any one moment.

Human services is also a procedure that is based on the previously stated philosophy. Concern for the whole individual as well as his environment leads to a "generalist" approach to training. As generalists, human service workers take responsibility for the whole person with all of his problems. As a generalist, the human service worker considers the client in regard to the various aspects of the helping process. If a specialist is needed, the human service worker is well acquainted with resources for help, and he makes certain that the person in need gets help.

In contrast to the human service generalist, the specialist concentrates on a particular kind of service. For instance, the specialist may be an ophthalmologist who examines, diagnoses, and treats a patient's eyes; a psychometrician who administers, scores, and interprets psychological tests; or a social worker who works only with groups of children in a community center. Another distinction is that the generalist is more likely to help fewer people with the larger view of all their problems. The specialist, on the other hand, cares for more people, considering more specific problems.

Defining human services is not easy. The phrase itself, "human services," takes in so much territory. A workable definition may be *human services refers to the attention and assistance offered to people by other helping people.* This broad definition covers a great many acts—from the neighborly chat to the professional person's aid. Before trying to pull together the meaning, an examination of the many possible human services is called for.

The many segments of human services

The centipede is a many-legged insect whose body is segmented (sectioned). Each part of the centipede's body has a pair of legs, which helps move that section. The important notion about the centipede is that although the pairs of legs may move separately, their movement is related to the entire body of the centipede. This interrelationship of the body to the legs is similar to the generalist idea (the body), with the legs representing the specialists. Both body and legs are essential for movement and change. Thus human services embraces the work accomplished by physicians, psychologists, social workers, mental health technologists, community service workers, and others. The help that these individuals offer takes in hundreds of different kinds of services.

The effectiveness of the generalist approach to helping people is dependent on a revolutionary change in educational procedures. In fact, human services demands social and economic as well as educational changes. All these facets of human conditions are interrelated. Social change requires "new economic relationships among human beings" (Hebert and Schulman, 1972) and the removal of the tracking system that the "A" to "F" or pass and fail grades encourage. The "place of education . . . becomes that of analyzing and fulfilling the needs of the people, by any means necessary." More education must be focused on providing services. More educators must realize that life experiences are learning opportunities that prepare individuals for human services. It appears as if the future will involve the student in the pursuit of lifelong learning both in and out of the school.

Roles and functions of human service workers

There are several barriers to the full utilization of nontraditionally trained human service workers. In addition to the notion that academic degrees are necessary for an effective helper, other obstructions hamper change.

Job-factoring approach versus developmental approach

The essential difference between the job-factoring approach and the developmental approach to role analysis is their centers of interest, that is, job tasks or the person to be served. Job factoring focuses on an analysis of the specific tasks and related activities that various levels of workers perform. This is the form upon which most of a person's work is decided. The worker's job is divided into functions, and the less desired, less responsible tasks are given to the individual with a lesser degree.

The developmental approach is different. This approach starts by exploring the needs and problems of the clients, their families, and their communities. Second, the tasks and activities that will satisfy these needs are determined. Third, similar tasks and activities are grouped into "job families." Finally, these tasks and activities are assigned to various levels of workers in accordance with three criteria as guidelines (Southern Regional Education Board, 1969):
- What the worker does with things (typewriters, other machines).
- What he does with data (" . . . data gathering, record keeping, data analyzing, and data synthesis").
- What he does with people (" . . . interviewing, coaching, counseling, teaching, treating").

These three criteria are evaluated in terms of the degree of responsibility involved and the difficulties that might be encountered with things, data, and particularly with people's problems.

Individuals at many levels of education and experience may be performing similar roles and functions. The difference in the jobs would be in terms of the closeness of supervision required, the extensiveness of responsibility and decision making, and the complexity of problems in which the various levels of human service workers may engage. For instance, the Southern Regional Education Board distinguishes the functions of the "outreach" role according to the functions at the four levels of responsibility (p. 6).

HOW IS A PROFESSIONAL SELF ESTABLISHED?

The stage is set. Human services provides the script. Who should be the actors, the

---

Outreach (detection, referral, follow-up)*

LEVEL I

- Do outreach visits, calls, etc. to homes, families, and neighborhoods to detect people with problems, to help them to understand the problem, and to motivate them to seek help. Let people know where help is available.
- Assess and decide how to best handle problem.
- Do outreach to follow up clients and assure that they are progressing with their rehabilitation in the community.
- *Make* self available—not just *be* available.
- Work with families at home or in office to help implement services and interpret laws, policies, and regulations.
- Interview and gather information.

LEVEL II

- Reach out to small groups (neighborhood groups) for detection of problems and understanding.
- Reach out to organize and follow up groups (alcoholics, ex-patients, offenders).
- Reach out to work with prisoners, the physically disabled, and others who can't come to mental health centers for services.

LEVEL III

- Reach out to community groups and agencies (orphanages, churches) to help them appreciate and manage psychosocial problems.

LEVEL IV

- Reach out to major agencies, industries, etc. to help them identify, analyze, and solve psychosocial problems (i.e., alcoholism, absenteeism).

---

*From Southern Regional Education Board (1969).

---

doers? What should these agents for social change be like?

Explaining the concept of the professional self

There are at least two directions that a definition of the term "professional" may take. One definition sets up a closed system in which individuals with degrees establish a monopoly of knowledge, skills, ethical procedures, and other credentials in order to regulate and maintain a certain status and levels of performance determined by an agency or by an accrediting group. Another definition provides a more open system in which individuals are presented with expectations and levels of training in knowledge, skills, and performance so that they may function in certain roles with specified levels of responsibility. This second definition follows the first in its stress on the obligations of a code of ethics responsive to individual rights and to society. The difference between the two definitions is in the acceptance of the nontraditionally trained, compe-

tent individual who has no degree as a beginning professional.

The other word, "self," focuses on the individual who is the actor or doer of the professional roles assigned to him. "Self" includes the thoughts, opinions, feelings, and strivings of which the actor and others are aware.

The *professional self* is therefore the result of certain general and/or specific training in knowledge and skills that produces an individual who meets established qualifications for certain levels of responsibility and competence. Furthermore, in the process of this training the individual develops certain attitudes toward his role as a helper that are related to the rights of the people he is helping and the helper's responsibility in preserving these rights (Ennis and Siegel, 1973).

A bucket of terms for levels of professional status

The word "nonprofessional" has been used most often to indicate the nontraditionally trained worker without a degree. This term is frequently applied as a catch-all for volunteers, parents trained to help their disturbed children (filial therapy), college students as play therapists, and mature women trained to be mental health counselors. Other terms are thrown into the same bucket with "nonprofessional."

paraprofessional According to Webster's dictionary, *para-* has several meanings—"alongside of, going beyond, like, resembling, or in a secondary capacity" (medical conception). Therefore paraprofessional would indicate an individual functioning either alongside of, or sometimes going beyond, the professional. He would act in a similar but, perhaps, secondary duty, which resembles the activities of the professional.

subprofessional The dictionary defines *sub-* as "under, beneath, below, lower in rank or importance, inferior or subordinate." Therefore a subprofessional would be an individual whose position is lower than the professional's and whose function is inferior and/or subordinate to the professional.

indigenous nonprofessional *Indigenous* means "native or produced naturally in a certain region or country." The indigenous nonprofessional is an individual born and living in the client's community, usually an inner-city, black ghetto.

functional professional This is a phrase coined by Carkhuff (1971) that describes individuals selected on the basis of their existing abilities in a given area who are trained to a high level of expertise in that area. "Although they do not have the formal or educational credentials they can function effectively as professionals." These individuals are usually indigenous (native) to the area from which the clients come for help.

beginning professional McPheeters of the Southern Regional Education Board has often used this phrase to describe 2-year community college graduates who are trained in human service programs, no matter what their programs are labeled. This phrase is the preferred title for this new brand of professional since it does not suggest inferiority or primarily assistant status. This phrase also allows for differences in levels of professional responsibilities. "Beginning professional" opens up the prospect for continued growth as well as for the utilization of existing skills from life and job experiences. Professionals would become even more important in training and supervision. In fact, supervision may be arranged so that aspects of training and supervision may be divided among the various levels of workers.

Characteristics of the professional self

There is one tool that is essential to human services—the professional self. The professional self is a defined identity that provides a social role with required levels of competence and opportunities for accomplishment in one's area of competence. This means that when an individual is relating to another individual in the role of a helper, he is persistently aware that he must be functioning at a higher level than the client and sharing himself for the client's personal growth. The helper may advance in self-understanding in the helping process but this is a side effect of the interpersonal relationship. The question of how this professional self is established remains. Two areas of information are probed.

Procedures to develop self-understanding

It would not be too farfetched to say that one may gather a great deal of information about oneself and yet have very little understanding. Understanding signifies full awareness, clear perception, and thorough knowledge. Understanding oneself may be a lifelong ideal rather than an immediate achievement to be realized. However, in order to penetrate the outer shell of information and attempt to get in touch with oneself, the individual should at least discover his learning style and identity-concept.

Learning style. People learn through their senses by interacting with an environment. This environment may be a book, a lecture or discussion, some physical contact with other people, or a combination of any or all of these channels for communication. Learning styles are conditioned in early childhood by means of the significant people (parents and others) around the child. For example, John thinks and learns best when he listens and then gets into a rapid and lively talk about things; Rose prefers reading and thinking before she responds to a question; Ingrid likes to learn by doing, touching, and interacting with others. John's learning style is auditory (hearing) and he probably is a quick thinker who is impulsive (dashes off, blurts out) in his responses. Rose is a visual (seeing) learner and is more cautious and slower (reflective) in her statements. Ingrid learns best through physical means (kinesthetic) and is more responsive to social learning situations (field dependent).

Knowing one's own learning style and being able to recognize learning styles in others is important for the educator as well as for the human service worker. Teaching as well as helping methods should be adjusted to the individual differences in learning styles of students and/or clients. Smith and Martinson (1971) conclude from a study of counselor-client interview relationships "... that counseling educators should consider the possibility that counselors' and counselees' learning styles have some influence on interview behavior. ... Counselors should be sensitive to their own learning styles as well as to those of their counselees and should learn to make whatever adjustments are necessary during the interview in order to facilitate a genuine therapeutic relationship."

Identity-concept. Although learning styles and identity-concept are presented separately, they are actually interrelated. In fact, individuals may indulge in certain characteristics in one setting and other aspects in another setting. However, each individual seems to maintain a general pattern of learning style and identity-concept that is consistent and persists.

Identity-concept relates to the unique ideas an individual has about his role and status in society. This concept would include thoughts about maleness/femaleness, bright/dull, educated/noneducated, professional/nonprofessional, and so on. The following brief exercises focus on the search for identity.

Exercise 1-1
GETTING AT THE "WHO" OF YOU

You need a pencil and thirty small pieces of paper separated into three piles of ten each for this exercise. Answer the following questions in the given order and write one answer on each of the ten slips of paper.

1. Who am I?
2. Who do others think I am?
3. Who would I like to be?

Write anything you think about in answer to the questions, even if the replies seem silly to you. It is important that you write your thoughts as rapidly as you can. Turn over each answer as you complete it. Do not look at your earlier answers. Just keep on writing until you have ten answers on the separate sheets of paper for each question.

After completing the answers, write on a separate piece of paper and/or discuss with other group members what the answers you have written tell you about yourself. Compare, also, differences in the answers to questions 1 and 2, questions 1 and 3, and among all three sets of answers.

Exercise 1-2
WHAT'S IN A NAME

Complete the following four steps before proceeding to the further instructions. After you complete each step, rapidly write down any thoughts that came to you as you were writing.

1. Write your first name *very* slowly.
2. Write your middle name *very* slowly. (Omit this item if you have no middle name.)
3. Write down your last name *very* slowly.
4. What would you like your name to be—first name, middle name, and last name?

When you complete steps 1 through 4, get into smaller groups and talk about your responses. Concentrate on what you found out from the written exercise and what you further discover about yourself and others during your discussion.

Exercise 1-3
OBJECTS AND PEOPLE

Arrange the group into a large circle. Look around the room at the other group members and at the objects in the room that either you or other people have. Pick an object or a person that seems to express what you are or what you would like to be. Write down the name of the person or the object and briefly explain what about the person or object expresses *you* and/or what you would like to be. Then discuss what you wrote with the other group members. Explain what made you prefer a person to an object or an object rather than a person to represent you.

Exercise 1-4
UNMASKING

Think about the following questions and then answer them in terms of most of your experiences and responses to people.

1. Do you feel you are an open person?
   Yes ____ No ____ Sometimes ____
2. Would you prefer that people do not know too much about you?
   Yes ____ No ____ Sometimes ____
3. Do you feel free to tell people about yourself?
   Yes ____ No ____ Sometimes ____
4. Would you get upset if someone is told something about you that you prefer they did not know?
   Yes ____ No ____ Sometimes ____
5. Do you like people to ask you questions about yourself?
   Yes ____ No ____ Sometimes ____
6. Do you feel exposed when people seem to know too much about you?
   Yes ____ No ____ Sometimes ____
7. Have other people told you that you are frank?
   Yes ____ No ____ Sometimes ____
8. Do you like people to know you better?
   Yes ____ No ____ Sometimes ____
9. Would you feel comfortable in writing about your life experiences?
   Yes ____ No ____ Sometimes ____
10. Do you feel people are prying if they ask you questions about yourself?
    Yes ____ No ____ Sometimes ____

After completing these questions, look over your answers. What do you see as the predominant trend in your answers?

Although all aspects of the concept of identity are not examined in these exercises, the characteristics that are explored help define some of the boundaries of one's feelings about oneself—about one's place in the world around him. Who a person thinks he is reflects the more important elements of an individual's identity-concept. These essential characteristics may be the person's role on the job, at home, in the educational institution, or as a man or a woman. If a vast difference is demonstrated between who the individual thinks he is, who he thinks others think he is, and who he would like to be, this person may be revealing his dissatisfaction with himself as he *is*. On the other hand, he may be indicating that he places a phony front between himself and other people.

A person's name is one of the trademarks of identity. Thus, if the individual has negative feelings about his name, it is likely that he also carries around negative feelings about his identity. Still another charac-

teristic of identity is the preference an individual has for either objects or people. This point is also discussed under learning styles. The choice of person or object tells something about an individual's interpersonal inclinations. The explanation of the basis for the choice of a person or an object further aids in understanding how the person feels about himself. Finally, how much an individual opens up to others and the specific choice of others to whom the person discloses himself is an indication of the degree of comfort an individual feels in sharing his ideas and feelings with others.

A general summary description of the characteristics of the individual likely to be an effective counselor would be: The effective counselor perceives his role as a helper high on his list of "who-ness"; he also perceives others as seeing him as a helping person. He is comfortable with himself and with his name, although this does not signify that he completely likes himself or his name. People rather than things catch his interest and concentration. Finally, the effective counselor is able to comfortably share himself with others, not for boasting, but instead to further a closer relationship (intimacy).

A code of ethics

The preamble (introduction) to the American Psychological Association's Ethical Standards for Psychologists (1967) conveys the essence of the meaning of ethics.

The psychologist believes in the dignity and worth of the individual human being. He is committed to increasing man's understanding of himself and others. While pursuing this endeavor, he protects the welfare of any person who may seek his service or of any subject, human or animal, that may be the object of his study. He does not use his professional position or relationships, nor does he knowingly permit his own services to be used by others, for purposes inconsistent with these values. While demanding for himself freedom of inquiry and communication, he accepts the responsibility this freedom confers: for competence where he claims it, for objectivity in the report of his findings, and for consideration of the best interests of his colleagues and of society.

The APA statement supports the notion that ethics are more than "do's" and "don'ts." In addition to setting up regulations for ethical behavior, this preamble offers certain values as part of a philosophy about the people with whom the helper becomes involved. The following items should be carefully considered and become part of the human service worker's professional self.

competence Refers to the capabilities and limitations in knowledge and skills that exist at the time when a client comes to the human service worker. The helper must be able to recognize as well as admit how much he does not know and how much he does know.

referral Implies that when the helper discovers that he is unable to assist the client any further or when the client asks for a change of helper, then it is the helper's responsibility to seek out and refer the client to someone else.

privileged communication Stresses that the human service worker's direct line of responsibility is to his client first rather than to the educational or other institution in which he is employed. The helper may not discuss his client's communications or release information about him without the client's permission. Any deviation from this practice of confidentiality must be thoroughly considered in the light of its risk to the client. Information may be revealed " . . . only after much careful deliberation and when there is clear and imminent danger to an individual or to society, and then only to appropriate professional workers or public authorities" (APA, 1967).

right to treatment This is a two-edged statement. The client has the right to demand help and to receive adequate treatment in a public institution or from privately paid services. The client also has the right to refuse a particular kind of treatment and to seek out another form of help if he so desires.

privacy Involves respect for the individual. Herding individuals into showers; forcing them to use toilets without walls; invading the client's space without his permission; talking about the client to someone else in the client's presence; and testing, researching, or taping him without his permission all show low regard for the dignity of the individual and also are unethical.

These are some of the more essential items. Trainees in human service programs would be wise to discuss these points and arrive at an even larger list of specific items related to a code of ethics.

CHAPTER TWO
LEARNING TO OBSERVE, RECORD, AND REPORT

Learning to observe. (Photo by Ruth Silverman, Editorial Photocolor Archives.)

There is one ability that the human service worker must have in large quantity—the ability to observe. Everyone observes from the moment of birth. This is the route for learning. The infant attends to his mother; he watches and listens. On the basis of observations the infant, the child, the young adult, and the older adult get the information they require in order to respond to the world around them. During his development the infant picks up some quirks that affects the accuracy of his observations. Perhaps the infant thinks, "Momma is good; she has blonde hair; she does not wear glasses; she has a high-pitched voice." The associations he makes with Momma—her hair, her eyes, her voice—reinforce certain ways of seeing, hearing, and relating with other women. Observations are tainted by the experiences an individual has with other people. Thus all people observe other people; yet few people observe without being influenced (contaminated) by their past experiences. The individual who is working with people must decontaminate himself of these influences before he can improve his ability to observe.

This section is about observation and what is done with observations so that they make sense to the observer as well as to others who may become involved with the client.

OBSERVATION

Everyone with intact sensory apparatus believes he sees, hears, touches, tastes, smells, and gains knowledge about the world through these "doors to perception." The important consideration is what kind of knowledge is learned. All observation is the process of obtaining facts or data through one's senses. The difference comes from the degree of orderliness, planning, and awareness of contaminating influences. When the act of observation and the qualifications for accuracy are combined, observation can be defined as the cautiously planned, orderly act or process of fact-gathering through one's senses.

The importance of learning to observe

The three monkeys huddled together are examples of the opposite of observation. Like the proverbial ostrich who hides his head in the sand, the three monkeys "see no evil, hear no evil, and speak no evil."

The monkeys' kind of nonobservation is what the effective human service worker must avoid. Watching, listening, and appropriately using the information are the routes to take to understand behavior. The human service worker must be able to answer three questions about his observations.

What for?

To know "what for" is to know the purpose of the observation. If an individual is to be evaluated to determine, for example, his qualifications for a job, the observations focus on the specific education, experience, and skills needed for the job. This may be accomplished by means of an interview, tests, a questionnaire, or other methods that provide facts about the level of the individual's competency. Observational procedures for the job-seeker are much less time-consuming than for the individual who is troubled and requires more extensive interviewing/counseling or psychotherapy. Diagnosing and planning objectives to be accomplished demand more observational time and a larger variety of techniques. Yet observation is the basis for the decisions about the job-seeker as well as for the individual with problems. "What for" merely gives direction to the specific procedures to be used but does not alter the basic procedure of observation.

How?

The "how" of observation is to some extent dependent on the "what for." "How" also requires an understanding of the differences between observations (facts) and inferences (guesses, interpretations). The "what for" sets the purposes, the "how" decides the procedure of fact-gathering, and the "what" refers to the specific points to look for when observing.

What?

Most important to thorough observation is "what" is observed. Each person sees, hears, and feels *selectively;* that is, each person's reality is colored by the way he looks at things. This leads to selective inaccuracy. Anyone who has played the game "Gossip," or whatever else it is called, knows how a sentence can be altered when it is whispered from one person to another around a circle. This is how rumors are born and grow with added twists as the message is passed from one gossiper to another. Anyone who has been a witness in court soon becomes aware of the variety of descriptions that can be given for the same event. The observer of human behavior must therefore develop his skill in objective reporting. "What" calls to attention possible slanting of observations according to the observer's pet likes and dislikes. "What" also asks for a detective-like search for cues (clues): "talking cues" and "doing cues." These two types of cues are discussed later in the section on developing observational skills.

Factors that influence accuracy of observation

Awareness of the "what for," the "how," and the "what" of observation sets the stage. Consideration of the factors that influence the accuracy of observation places on the stage the more and less obvious interferences to exact observations. Broadly speaking, there are two interrelated areas that influence the way in which an individual observes. These are the physical and psychoecological areas.

Physical factors

Everybody believes he is alert to the physical influence. A more careful examination of the physical aspect, however, reveals some potential blunders that are often hidden.

Degree of sensory acuity. When the individual begins to view a world made fuzzy by nearsighted eyes, he is very likely to get corrective lenses to sharpen his vision. However, the fuzziness of sounds is not as readily detected. It is so easy to fill in with one's own words what one does not hear. The special inflections (changes in voice tones) also may be lost when the individual with a slight hearing loss strains to hear the main parts of the message. Not only the sharpness of one's senses (sensory acuity) but also the position from which one is observing may influence what one observes. One more factor that applies to the use of an individual's eyes and ears is the way in which he attends to what is happening around him. There are the more observant and the less observant and the many individuals in between these two poles. Later, attending behavior, that is, listening and

looking, is discussed at greater length. The following example demonstrates the different observational results due to different degrees of sharpness in hearing.

Estelle says, "Marty was busted for selling dope." This is a statement of fact with no specific changes in tone and no emphasis on particular words. However, Estelle might have emphasized the words "Marty," "busted," or "selling," and the sharper hearing of the more alert observer would hear, "*Marty* was *busted* for *selling* dope." In the first statement, Estelle appears to accept the possibility that Marty might get busted and also that he might be selling dope. In the second statement, with the emphasis on "Marty," "busted," and "selling," Estelle shows her surprise and shock that Marty, of all people, has been arrested for selling. The one sentence Estelle utters may have different meanings depending on the loudness or softness of the words and the rising or lowering of the voice tone. The keen observer notes these characteristics if his hearing and attending are adequate.

The words themselves have different meanings depending on the individual's background. Estelle's statement is one example of this special meaning for words. "Busted" for some individuals may carry another message rather than being arrested. For some people "busted" refers to the well-proportioned, ample bosom of a woman. Another example of the need to not only hear clearly but also to understand the meaning of the words and how they are being used is in the statement, "That's really bad." The tone of voice, the loudness or softness, and the person making the statement are taken into consideration. Thus this one sentence might be telling someone, "I do like your faded jeans" or "That must have been interesting or exciting" or "Sorry to hear that; must be unpleasant." The point is that accurate hearing is one part of accurate observing.

State of health. An aspect that is related to sensory acuity is the individual's state of health. The way an individual physically feels has a great deal to do with how sharply he is able to as well as how much he wants to look and listen.

Fatigue, a clogged nose, an itch on one's arm, all may have an effect on the efficiency of an individual's observations. How interested might an individual be in observation if he is experiencing hunger pangs. Is it possible to be exact and complete in one's observations while feeling physically uncomfortable?

These points are not stated to stop individuals from observing under less than ideal physical circumstances. Instead, they are mentioned to alert the observer to the dangers involved in depending on only one brief observation. These physical factors are often more obvious than those that stem from interpersonal relationships and environmental influences, the psychoecological factors.

Psychoecological factors

The psychoecological label places in one bag a number of variables (factors). Psychoecology is made up from two words, "psycho" from psychology and ecology. Psychology is the scientific study of the behavior of living organisms. This leads to a study of the individual's feelings about himself and about his relationships and transactions with people. Ecology, particularly human ecology, is the examination of how an individual handles transactions with his environment. This viewpoint encourages the observer to develop an increased awareness of the numerous factors that enter into human transactions. Psychoecology brings into focus the interdependence of all the people, animals, things, physical surroundings, and so on in an individual's behavior.

Situation/location of the observation. The starting point for the observation of an individual is what is happening around him at that particular moment. Other people present just prior to or during the observation as well as the weather, the temperature, and the atmosphere inside and outside the room are part of the observational picture. Not all of these factors may have equal significance. Yet some of them may be manipulating the observed person to respond in a certain way. A room that is too hot or too cold will affect behavior. These facts should be noted so that later observations may be compared if temperature changes do occur. The following comparison of observations of John's behavior on two different occasions discloses some interesting differences.

1

John is seated at the table next to Sarah, who is busily working with clay. Four other 5-year-old children are also seated around the table; each of them is working with something—puzzles, clay, or crayons. John just sits and moves his head and eyes from one person to the next. He turns his chair to face the observer. In moving his chair around he bangs into Henry, who is seated next to him. Henry's puzzle part falls. With one movement of his hands, Henry picks up the puzzle and throws it toward John. John, running toward the observer, is not hit by the flying pieces. John does not stop at the observer but continues on to the window and looks out. The sky is getting darker and a slow rumble seems to be coming from the distance. John runs toward the observer, touches her knee with his tightened fist, and then continues to the teacher, who is seated in a circle with another group of four children reading them a story. John runs to her and falls across her lap with his head and arms hanging over one side and his legs over the other.

2

The next day John is in the housekeeping corner digging in the large box of "big people" clothes. He finds a bright red, striped tie and a policeman's hat. He puts the tie around his neck and loops one side over the other. Then he pulls the policeman's hat on the back of his head. Mary, who is in the corner with him, calls out, "Bwekfuss." John walks over to the table and picks up a cup, putting it to his mouth while he bends his head back. He "drinks" and then goes over to Mary and plants his lips on Mary's cheek, making a loud smacking sound with his lips. Then he goes over to the window and looks out at the brightly sunlit small hill and the large green field surrounding the school. A smile spreads across his face and he turns toward the observer and says, "Hi."

• • •

Same child, different weather. The meaning of the difference in behavior is not interpreted by these two incidents. However, John in the first observation did act differently than in the second observation.

Further facts are needed to better understand which pattern of behavior is more likely to recur.

Individual characteristics of the observer. The observer's degree of awareness, his flexibility, and his life experiences regulate the thoroughness of his observations. The basic question is, "How open is the observer to his own life space and to the life spaces of the people he observes?" Life space refers to all the factors that influence the behavior of a person at any moment. These factors include objects, persons, goals, thoughts, and other environmental events. The more aware observer knows the many complex transactions going on at the time of his observation. This observer heeds basic differences in language usage and understands that words mean different things to different people (semantics). In addition, he painstakingly sifts through his prejudices to work at lessening their effect on his observations. The alert observer knows how biases against anyone because of race, religion, or appearance will shut him into a tight box of inaccurate and incomplete observations. The biased individual sees what he expects to see, hears what he thinks should be said, and is more likely to observe something happening because he expects it to happen. Observations by such an observer are fixed according to his prejudices. Even the observations of an observer who is too biased in favor of the person he is observing can pigeonhole his thinking.

In the following example Mr. T.'s observations contrast with Mr. B.'s observations of the same event. Which observer is more prejudiced?

1

Mr. T. observes Sonia, who is 72 years old. She is sitting in a chair with her hands lying in her lap, palms up. Her eyes are on her fingers as she moves them one by one. She moves the thumb of her left hand several times and then a smile slowly spreads across her face. Mr. T. describes her smile as "broad and silly looking." He writes, "Her smile spreads over her face, cracking the wrinkles at the sides of her mouth. She continues to smile while she moves her left thumb. Then she begins to move her left leg up and down, jiggling her hand in the process. Her left shoulder moves slowly,

then more and more rapidly. She looks toward her shoulder and watches it as she gradually increases the speed of movement." Mr. T. adds the comment, " . . . obvious senility, possible feeling of being followed, silly behavior that shows lack of contact with reality." He also thinks, "Just like my grandmother when she got old. What a burden she was!"

2

Mr. B. observes Sonia but does not write that her smile is "silly." He also omits comments about Sonia's "obvious senility, feeling of being followed, silly behavior that shows lack of contact with reality." Instead, Mr. B. writes, "Something about this woman is very appealing. Wonder what she is thinking about as she moves her fingers, her leg, and her shoulders? Must find out about possible arthritis. Perhaps she smiles because she is so pleased that she can move her fingers."

• • •

How open are Mr. T. and Mr. B. to their own and to Sonia's life spaces?

Life experiences are another characteristic that enters into the observational process. How extensive and varied is the observer's experience? Prior experience with a variety of people in many situations is more likely to lead to openness of observing, openness in seeing relationships, and greater flexibility in searching for cues. In addition, educational experiences that prepare the observer with developmental information, knowledge of needs, and awareness of behavior patterns also sharpen observational ability. Furthermore, life experiences are closely intertwined with the cultural background of the individual.

Cultural aspects. There is a simple truism, so simple that it is sometimes forgotten by an observer—"each individual acts differently depending on the other people around him." This statement points to the fact that when an individual is at home with his parents, wife, or children, he behaves accordingly. When the same individual is in his office, in the classroom, or in church, he acts differently. In other words, individuals behave in accordance with the expectancies of the social situation in which they find themselves. These social selves are dependent on the cultural differences in behavior the individual learns.

The observer has to realize this cultural effect. He must take into account not only the factors mentioned up to this point but also his and the observed person's social, economic, and ethnic background. In this sense the observer must realize that his presence, the color of his skin, the clothes he wears, and even the way he sits are influential on the behavior of the person he is observing. Both the observer and the observed person act differently when they are together than when they are apart because of their cultural differences. This is just another point about which the observer must be aware.

• • •

Awareness of all the factors that influence the accuracy of an observation does not magically do something about them. However, it is the beginning of learning to deal with possible distortion that might result from physical or psychoecological interferences.

The psychoecological approach of observation is particularly pertinent to the philosophy about people that is developed in this book. This model of thinking about people draws attention to the need to go beyond a one-shot, one-cause relationship between stimulating factors and resulting behavior. The helper who accepts this philosophy takes on added responsibilities to involve the home, neighborhood, school, and agencies in a coordinated plan to help the questioning and/or troubled individual. New patterns of interpersonal relationships are sought in a helping system that reaches out beyond the interview/counseling room. Many individuals become part of the helping team.

General observational procedures
Degree of involvement

Secondary sources. Instead of speaking directly to the person observed or watching his nonverbal behavior, the observer goes to files in which reports and other aspects of the case history are kept. Additional secondary sources that may be explored are other people who have had some contact with the person being observed. Thus the individual is observed indirectly by means of records, reports, or comments made by other people. This includes comments and observations written by other people as well as results of physical and psychological tests. Except for the tests, which are based on information acquired by means of some standardized measuring procedures, these impressions of others are likely to be based on prejudices for or against the observed person. These impressions should not be discounted even if they are biased since these ideas and feelings have had an effect, more or less, on the observed person's behavior. The next level of observation gets a little more into the act.

Spectator observation. The observer is still attempting to keep his presence from influencing the observed person. He attempts to place himself outside the focus of attention of the observed person. One way of doing this might be for the observer to sit out of the immediate range of vision of the observed person. Another method is to talk with someone other than the observed person far enough away so that the observer's conversation cannot be heard by the observed person. If the observer becomes part of the immediate activity of the observed person and still continues his observation, he involves himself much more in what the observed person is thinking and feeling. The observer also is adding another element, himself, which influences what the observed person does or says.

Participant observation. In this kind of observation the observer is actively involved with the observed person. He may be playing a game, reading a story, assisting in arts and crafts, or discussing something. At the same time the observer is noting what the observed person says or does. He is also mentally recording what is happening between the observed person and himself as well as with the other people and things in the surroundings. The remarks of both the observer and the observed person become an essential part of this observation.

Interview. The interview begins to wrap the observer and the observed person into an even closer net of care and concern. The interview may be brief and seek only quick, surface information for job placement or credit extension. The interview could be in depth to determine more about the thinking and particularly the feelings of the observed person. Another distinction in degrees of involvement also may be made between interviewing, counseling, and psychotherapy. Each of these transactions becomes more absorbing and more consuming of time and attention.

Type of observation

This classification is primarily for the convenience of discussion. The five-part pigeonholing of types of observation indicates differences in approach and in exactness. Yet the categories do overlap and one category may lead into another. For example, a random observation may become a carefully planned observation according to certain time intervals and later be set up as an experimental situation. The four degrees of involvement previously described may be used with the following five types of observation (Table 1).

Casual observation. Everybody indulges in casual observations that are based on a commonsense approach. Out of this kind of observation often emerges such statements as, "He's a brat," "She is happy," "She is hostile," and "He is a general nuisance." None of these labels actually describes behavior. Casual observation is hindered by past experiences, expectations, conventional labels, and the observer's ability to look and listen. The beliefs about himself and

about the people more familiar to him would interfere with the accuracy of the observation. An observation of someone known very well in familiar surroundings tends to be highly contaminated with feelings. It is much easier to observe the behavior of strangers than the behavior of close relatives, lovers, or friends. However, accurate observation is not a skill born with the individual. It must be learned and practiced.

Naturalistic observation. The aim of scientific observation is to gather precise and useful information about an individual. Information must be stated in such a way that it can be verified by another observer who is qualified to observe accurately. This may be partially accomplished by naturalistic observations, which are running records of observation that begin and end at any point of time. This on-the-spot record is the one used most frequently. The observer jots down his notes during the observation or waits until he leaves the presence of the observed person. If the observer takes this latter route, he must remember that the longer he waits to write down his notes, the more inaccurate his memory of his observations will be.

This running record contains observations written in an anecdotal or story form. Since the observer does not make any planned attempt to change the situation of the observed person, this kind of observation is considered to be naturalistic. Field observa-

Table 1. Types of observation

TYPE	EXAMPLES	DEGREE OF CONTROL
Casual	Observation of natural events—sky, trees, birds, other people	Observations are more or less undirected and are often inaccurate, contaminated by prejudices and past life experiences.
Naturalistic	Running record (log or diary) in clinic, classroom, or other individual or group situations; observations in the field; case studies	Observed events are uncontrolled. Observer exerts orderly approach and cautious observing. Observed person preferably should not be made aware that he is being observed.
Standardized	Time-interval observations and psychological tests	Conditions of observation are specified and must be followed. However, observed person is not controlled by changing any events around him.
Specific goal	Observation of specific behavior and for specific purpose	Observed events are carefully controlled in terms of events to be observed or ends observation is to serve.
Experimental	Laboratory study of specific kind of behavior	Control of variables is more or less extensive depending on what is being studied.

tions that explore the behavior of animals or of humans are an example of naturalistic observations. These observations may be written, photographed, filmed, or taped. Sometimes a combination of these methods is used. The disadvantage of naturalistic observations comes from substitution of imaginative inserts to fill out genuine observation. An observer who writes, "David looks at Cecily with hate in his eyes and hands," is recording his own wishes or bias about David and Cecily. A statement to the effect that David walks over to Cecily with his raised fist hints at a similar idea but does not provide unsupported conclusions about David's actions.

Clinical observations are a more structured (organized) example of naturalistic observation. These clinical studies often are referred to as case histories or mental status examinations. A case history is a scientific biography of an individual, an institution, or a group of people. The observer helps the client reconstruct his past history as well as his present complaints. All of these facts are pieced together in order to gain a fuller understanding of the forces affecting the individual's present problems. One disadvantage of the case-history method of observation is that contemporary observation depends on the observed person's memory. This may be a disadvantage in terms of the accuracy of the memory but does not detract from the value of observing how the individual responds to his memory. Clinical observation also includes the on-the-spot observation of the client during the time of the observations. Naturalistic observations often serve as exploratory studies from which standardized observations emerge.

Standardized observation. There are many forms that standardized observation may take. Two are time-interval observations and psychological tests. Both of these varieties of standardizing procedure are based on obtaining a sample of behavior. Time-interval sampling requires observations at predetermined times of the day and usually for a set length of time. The length of time may be set from a few seconds or minutes to several hours. Sometimes the time samples may be half-hour or hour observations conducted at different times during the day. At other times the observation

may be in terms of the number of times a certain behavior occurred within a specified unit of time. For example, the question may be, "How many times during an hour does the child get out of his seat?" An observer who uses the word "constantly" to describe the frequency of such movement is providing inaccurate information since constantly means a different amount for different people.

Psychological tests are standardized (experimentally constructed and scaled) as to their administration and method of scoring. These tests provide a method to measure the observed person's abilities, interests, attitudes, and achievements. Although the rules for administration of the tests and the norms for comparison may be scientifically established, there are still two obvious factors that are not so easily controlled. The observer and the observed person have their own bags of behavior and the contents of these bags, the personal characteristics of each individual, may influence the outcome of a particular sample of behavior at a particular time. Therefore tests alone or any form of one-shot observation does not assure exactness of the observed person's lifestyle.

Specific goal observation. The observer may set specific goals in terms of observing specific behavior or in terms of a specific purpose for the observation. Both of these goals require precise planning. Observation of specific behavior may concentrate on verbal or on nonverbal cues. It may also focus on such specific acts as the cooperation of one child with another, the shifts of leadership within a group, or the frequency of the use of certain words. Specific purposes for observations may pose such questions as, "How accurate are the interviewer's skills of observation?" and "How does the individual control his emotions?" Specific behavior and specific purpose fuse into one another very often.

Experimental observation. The experimental observation is carefully planned and controlled. Consideration is given to variables and the method of measurement of the observations. A variable may be a drug, an event, or a person. Another word for variable is "factor." Variables are factors such as drugs, events, and people that exert some influence on the observed event or person.

When an observer sets up an experimental observation or design, he is searching to determine the relationship between certain variables. For example, he may study the effect of different amounts of marijuana on the memory of an individual. The procedure for observation would require that the observer think about the numerous variables that might affect memory. Such factors as age, sex, situation, and past experiences are considered. The observer controls as many of the variables as he can and/or thinks necessary for the exactness of his experiment. He controls the age variable by selecting only a certain age group. He arranges the setting so that unwanted noises and other distractions are not present. He selects individuals who have or have not used marijuana, depending on the effects he wants to study. In other words, the observer controls certain variables and manipulates the independent variable (antecedent events), in this case, the amount of marijuana. He changes the amount in some predetermined manner so that he may observe the memory effect of different quantities of marijuana. The memory effect is the dependent variable (the consequent event), which is the change in the observed person's memory behavior. This is called a single-variable experimental observation.

Training techniques

Training techniques are employed in teaching the human service worker to use keen observation in order to understand better. Reaching a better understanding is a three-step process of (1) gathering information through direct observation of an individual and indirectly from comments and observations from other people, (2) thinking of hunches (interpretations, inferences, hypotheses) to explain the reasons (causes) for the observed person's behavior, and (3) testing hunches by determining whether there are enough observational facts to support the hunch. Additional observations may be needed to support or deny (refute) the correctness of the hunch.

Developing observational skills

Before getting into the actual practice of observational skills, a basic distinction must be made between verbal and nonverbal cues.

Verbal and nonverbal cues

" . . . One can observe without interviewing but one cannot interview without observing" (Kadushin, 1972). This statement refers to the "looking," "listening," "smelling," "touching," and "tasting" channels for observational cues. Observation, therefore, must be extended beyond merely word-listening cues. The effective observer is tuned in to the multichannels of communication—eyes (visual), ears (auditory), nose (olfactory), hands (tactile, touching), body, arms, and legs.

For purposes of convenience the messages that are sent through the communication channels are roughly divided into verbal and nonverbal messages. The verbal messages include both the particular speech patterns used and the nonword punctuation marks called paralinguistics. Nonverbal messages are even more extensive since these include the language of the body, parts of the body, and facial gestures. Figs. 2-1 and 2-2 spell out these channels for sending messages.

Verbal cues. These messages tell the observer a great deal about the observed person. The speaker's choice of words can reveal his socioeconomic status, his ethnic group, his religious orientation, his subgroup, his feelings about himself, and his feelings about other people. The language used by a less-read and/or less-educated individual reflects the social and economic setting to which he has been exposed. The geographical area he is from will be reflected in his words. Although the poor individual (both economically and educationally) may appear to have verbal inadequacies and difficulty in verbal expression, this is an inaccurate conclusion. Children from slum areas use words differently. They use a more informal, "public" language with simple grammatical construction and frequent use of such conjunctions as "so," "then," and "because." The precision of their words seems inadequate to middle-class listeners since they do not use "school words." Instead, their words are open to more imaginative expression because they are less word-bound than the middle-class individual (Deutsch, 1964).

The observer who writes, "poor vocabulary," to describe an individual from a slum area is not listening very carefully to the vast potential beyond the individual's words. He also is not watching the rich gestural language this "deprived" individual is demonstrating. Ethnic group language often is intertwined with the speech mannerisms characteristic of a particular socioeconomic status. Such words as "macho" (manly), "chutzpah" (shameless, impudent bravery), and "soul food" (chitterlings, corn bread, turnip greens) are now included in the dictionary. However, Mexican-Americans, Jewish persons, and black persons originated these words.

Religious orientation also insinuates itself into the words used. The language of the person close to his church or to certain religious convictions is sprinkled with words referring to the deity or to some other words such as "guru" (a spiritual teacher of the Hindu). Sometimes the words lose their religious significance or original religious attachments and become part of a culture, such as the golden rule ("Do unto others as you would have them do unto you") or part of a subculture. Many young people who have turned to oriental philosophy as their guide to living speak of the Hindu concept of Karuna, meaning the cultivation of wisdom and compassion for other human beings, which is, incidentally, an excellent guideline for the human service worker.

The language of subgroups varies from the strung-out, hung-up language of the counterculture to the middle-class "bookish" speech of the teacher to the self-centered speech of the aging. Each of these groups has its own vocabulary, which is not only similar for the group but also is individualized. For the youth group, the black group, and the drug group, the words used and their meanings change so rapidly that a dictionary published today would change within the hour of its publication. There are some words that remain such as "pad," room; "dig," to understand; "jazz," worthless talk; "cool it," take it easy; "bread," money; "knocked up," made pregnant; "skinpopping," injecting a needle under the skin; "nickel bag," single packet or dose of a drug costing 5 dollars. Some so-called ordinary words have been given new meanings. "Trip," for instance, signifies one thing to the "square" for whom it suggests travel by car or other vehicle. For the drug freak, however, this word means a self-trip induced by a psychoactive drug, which may

be a "bummer" (unpleasant or frightening) or a "slide" (a beautiful, glorious, or even spiritual experience). The list of subgroup words is not by any means exhausted.

For the older person as well as for the young child, feelings about himself are revealed by his words. The aging person and the child use self-centered (egocentric) speech. They talk to themselves and carry on monologues. Sometimes two monologues are conducted between two aging people in which each one is simply talking to hear himself talk, just as the young infant babbles and mouths words. For the young child this egocentric speech serves as a device for learning; for the aging person it is often a sign of his loneliness and depression. Words can reveal much about how an individual feels about himself and about other people. The observer should keep tab of the word choice. Does the observed person use words of certainty such as "I can" or words of doubt such as "If I could"? How close does the observed person seem to be to people? Is the observed person using words such as "I" and "we" or "they" and "you"? The following two statements indicate different degrees of warmth or closeness—"It seems inevitable that your needs will be met" and "We'll work on getting what you need." Which conveys the warmer message?

Qualifying words and interpretive versus descriptive words are further instances of the use and abuse of words. Qualifying words manage to take something away from the meaning of the sentence either by adding a statement or by suggesting that the action implied is unlikely to occur. "But" is one of these qualifying words. "You had a good grade on that test *but* you might have done even better." "If only" is another teasing insertion. "*If only* I had the time, I would help you." How often does the observed person, and the observer, too, use these words? Finally, perhaps the underlying idea for this lengthy discussion about words is the difference between interpretive words and descriptive words. Interpretive words evaluate and pass judgments; they label. Descriptive words explain what is seen or heard in behavioral terms, unblemished by personal bias as much as possible.

"She's an authoritarian." "She shouts her orders and repeats them even louder until you do what she asks."

"He is happy." "He jumped up and down."

In the first set of two sentences, "authoritarian" may fit the behavior. However, even in this situation, more would have to be known about the interpersonal transactions before a label is attached to the behavior. In the second two sentences the jumping person may or may not have been happy. He might even be showing the overactive (hyperkinetic) symptoms of the brain-damaged individual. In both sets of sentences the first sentence infers, interprets, and hypothesizes; the second sentence describes behavior. This distinction is particularly important when the observer records the event he is examining. A much-labeled account says very little about the observed person and much more about the observer.

The voice sounds and other characteristics that co-occur (are uttered at the same time) as the words serve to modify, emphasize, and regulate the message or the message-receiver. When the message-receiver is bombed by *"cut the crap"* from a voice that shifts rapidly from a slow, soft bass to a fast, loud screech, the message is plain. Pitch (loud or soft, high or low), pronunciation (clarity, style), and speed of speaking are some of these paralinguistic clues that provide the question mark or the exclamation point for the words used. A more complete list of other characteristics is included in Fig. 2-1.

It would be impossible to pick up all of the verbal cues listed. However, it *is* possible to become more aware of the words and sounds with practice. Exercise 2-1 provides a way to learn by doing.

Exercise 2-1
LISTENING TO THE WORDS
AND TO THE SOUNDS

Two trainees volunteer for this exercise. (Did you notice how positive that last statement is?)

The remainder of the group get into a circle and select different parts of the list of characteristics in Fig. 2-1 on which each person will concentrate. Thus some trainees concentrate on *word choice;* others select *pitch* or *speed.* After the selections are made so that all of the characteristics are ob-

served, the two volunteers get into the center of the group.

The group members turn their chairs around so that they are facing away from the two volunteers. This is done so that they may concentrate on words and sounds and not be distracted or influenced by the nonverbal behavior of the volunteers.

The group members use the list of characteristics in Fig. 2-1 for which they are responsible. Each person tallies (place marks next to) the characteristic each time they hear the observed person speak or voice it.

For 10 minutes the volunteers talk about some topic. During the first 5 minutes they agree about the topic. Then they talk for 5 minutes but this time they disagree, no matter what is being said.

The volunteers return to the group. Everyone turns around and a general discussion follows that focuses on the differences in the tallies between the 5 minutes of agreement and the 5 minutes of disagreement.

Nonverbal cues. If a dictionary of nonverbal cues could be compiled, it would be very helpful. However, except for a comparatively small number of nonverbal cues, the meaning of the cue is dependent on the time, the place, and the person using it. This need for specificity does not detract from the importance of the cue but does urge the observer to be concrete in his observation and recording. Nonverbal behavior embraces many different languages.

Body language. This includes the study of the distance between people and the orientation or direction of the body (proxemics) as well as the motion of the body (kinesics).

Gestural language. These motions include pantomimic and nonpantomimic gestures of the hands and fingers. *Pantomimic* gestures are actions that substitute gestures for words. The game of charades is an example. *Nonpantomimic* gestures are actions that accompany words and modify or regulate the meaning of the words. Some pantomime is formal and has specific meanings that are known by most people in a particular culture—waving good-bye, the "V" for victory, making two wavy lines in the air to demonstrate the mountains and valleys of the well-proportioned female, putting one's thumb to the nose with the remainder

	TALLIES				TALLIES	
Word choice			Louder			
Think			Softer			
Feel			**Pronunciation**			
Can			Clear			
Know			Slurred			
May			"Put on"			
Ought to			"Bookish"			
Must			**Speed**			
Should			Slow			
Would, if			Rapid			
Distance			Jerky			
You			Smooth			
They			**Voice control**			
It			Voice breaks			
We			Trembles			
Qualifying words			Chokes			
If only			Clears throat			
But			Smooth control			
Others (specify)			**Sound of voice**			
Interpretive words			Nasal			
Cooperative			Hoarse			
Elated			Others (specify)			
Troublemaker			**Speech sounds**			
Normal			Ummm			
Others (specify)			Ah			
Descriptive words			Hmmm			
Look			Mmmhmmm			
Smile			Uh huh			
Jump			Huh			
Walk			**Breathing**			
Frown			Deep			
Others (specify)			Rapid			
Pitch			Even			
Rises			Sighs			
Falls			Laughs			

Fig. 2-1. Verbal cues.

of the fingers extended to signify extreme distaste and annoyance, forming a circle with the index finger and thumb with the remainder of the fingers extended to signify "okay" or "going well," and making a sign of a circle next to the temple to represent "he's nuts." Subgroups also have their formal pantomimic language—the hand language of the deaf mutes, the military salute, the gesture of the hands in prayer, the clenched fist salute and the handshake of soul brothers, and the thumb up to save the bull and thumb down to kill him at the bullfight. There are also informal gestures whose meanings depend on the situation and the context (remainder of the content). This gesture may involve parts of the face that are touched by the hands or fingers such as pulling the earlobe, rubbing one's nose, adjusting clothes, squeezing a pimple on the face, and other characteristics mentioned in Fig. 2-2.

Nonpantomimic gestures point, chop the air, turn the palm, and so on. Pointing movements draw attention more specifically to a particular person or object—saying "that girl" with the index finger pointing toward her makes certain that the particular girl is noticed. Palms may be turned up in uncertainty or down or facing toward the message-receiver when being emphatic. "Cut the static" and "Don't interrupt" are both more commanding when a palm faces the message-receiver in a "stop" sign. Circular movements may demonstrate the generality of a statement ("Sometimes this works and the . . .") and semicircular movements of the hand back and forth suggest "either/or." Chopping gestures are used for emphasis and may be used to emphasize one word over another in a sentence. "You have to think about verbal and nonverbal cues" may be accompanied by two hand chops that emphasize that both cues are equally important or one chop after "verbal" or "nonverbal" to emphasize that one or the other is more important. There is the fish-story kind of gesture of expansion and contraction. Like an accordion, the size of the fish or whatever the object is made smaller or larger during the telling of the "fish story."

Facial language. The wrinkled brow, the smile, the blink of the eyes, and the gaze or stare of the eyes all regulate, communicate, and express ideas and feelings. Eye contact may be so intense that it annoys one or eye movement may start or stop a conversation. Wearing sunglasses makes the individual less available to others; this makes the removal of the sunglasses a significant sign of becoming available. On the other hand, the individual who wears corrective lenses may be removing himself from the world when he removes his lenses. Hands may be hidden in one's pocket, placed on one's lap, or sat upon. The legs and feet may be concealed behind a desk. The face is constantly exposed and, unless hidden behind a hand or mask, cannot conceal the feelings communicated. The interpretation of the meanings of facial language requires an alert, knowledgeable, empathic (keenly understanding) individual.

Language of objects. Personal adornments such as clothes, hairstyle, makeup, jewelry, and tattoos are expressive extensions of the self. This describes the significance of home furnishings also. There are some uniforms of group identification—for example, long hair, beard, headband, patched jeans, tie-dyed tee shirt, and beads portray a different life-style from the studded leather jacket (usually black), boots, and red, white, and blue helmet. Both of these attires contrast with an Afro haircut and dashiki or Bermuda shorts with an Indian madras shirt. The picture of the home of each of these people would differ also. Books, hi-fi records, plants, art on the wall, and the furniture style also express the life-style of the individual. These nonverbal cues from objects are part of the script that helps the interviewer gain better understanding of his client.

The more specific list in Fig. 2-2 alerts the observer to the many signs with which nonverbal cues "speak."

Exercise 2-2
WATCHING MOVEMENTS AND GESTURES

Two volunteers are needed for each of the three parts of this exercise. The remainder of the group selects different parts of the list of characteristics in Fig. 2-2 on which they concentrate. Some trainees focus on *posture*, others on *feet*, and so on. The number of different assignments depends on the size of the group.

	TALLIES			TALLIES	
Posture			Others (specify)		
Stiff and rigid			**Hands**		
Bent			Rubs eyes		
Loose, relaxed			Chin on hand		
Movement			Hands and arms drop over back of chair		
Bends forward					
Bends backward			Sits on hands		
Shrugs shoulders			Hand(s) on hip(s)		
Orientation			Hides hands behind desk		
Sits facing			**Palms**		
Sits sideways			Up		
Distance			Down		
Sits closer			Out		
Sits farther away			**Circular movement**		
Accessibility			Slow, continuous		
Open (arms and hands loosely placed on lap, legs uncrossed)			Semicircular, back and forth		
			Swing an arc		
Closed (arms held crossed across chest, legs crossed near thigh)			**Chopping gestures**		
			Rhythmic (up and down)		
Feet			Nonrhythmic (emphasizes certain words)		
Tap					
Kick			**Expansion/contraction waves**		
Shuffle			"Good-bye" motion		
Slide back and forth			Wavy lines		
Legs			Others (specify)		
Sway in circular motion			**Handshake**		
Kick back and forth			Firm		
Move up and down at bent knee			Flabby		
Slaps thigh			**Head**		
Hands clasp knees			Nods		
Cross one knee over other			Bobs (shakes up and down)		
Cross ankle of one leg over knee of other			**Hair**		
			Runs fingers through (combing motion)		
Arms					
Crossed at waist			Plays with		
Akimbo (elbows bent outward in semicircle or arc) on chest			Curls hair around fingers		
			Chin		
Odor			Thrust forward		
Perfume			Drawn in		

Fig. 2-2. Nonverbal cues. (Adapted from Wiener et al., 1972.)

	TALLIES			TALLIES	
Quivers (trembles)			Bad		
Neck			Smell of liquor		
Tense muscles			**Eyes**		
Stiff, rigid			Wide open		
Rapid hand-to-neck movement			Narrow slits		
Forehead			**Eye contact**		
Wrinkles			Look toward directly		
Furrows			Look away, shifts gaze		
Eyebrows			Wears sunglasses		
Lifted (arch)			Removes sunglasses		
Contract (pull together)			Removes corrective lenses		
Frown			**Eyelids**		
Nostrils			Close slowly		
Open wide (flared)			Flutter rapidly		
Lips			One or both wink		
Smile			**Fingers**		
Curl			Point		
Tremble			Drum on table		
Turn up			Extend		
Turn down			Pick things (lint) from clothing		
Open			Scratch		
Close			Pull earlobe		
Moistened by tongue			Rub nose		
Jaws			Squeeze facial tissue, pimple on face		
Clamp			Snap pencil		
Teeth grind			**Adjust** (straighten, pull)		
Face			Clothes		
Hides face behind hands			Things on desk		
Cover mouth or eyes with hands			**Rubs** (desk, arms, chair)		
Beads of sweat on face			**Pound table**		
Breath					

Fig. 2-2, cont'd. For legend see opposite page.

After selections are made and all of the characteristics in Fig. 2-2 are covered, the two volunteers get into the center of the group. The group members face the two volunteers in a semicircle. (If the group is too large to see the volunteers, smaller groups may be formed with several pairs of volunteers.) Group members use the list of characteristics in Fig. 2-2 for which they are responsible and tally the characteristics as they see them performed (emitted) by the trainee-volunteers. The sequence of the three parts may be changed in accordance with group interests.

Nonverbal behavior with known and unknown subject

One trainee-volunteer discusses a subject he knows well for 5 minutes. Nonverbal behavior is tallied by the group members. Then the speaker talks for 5 minutes about a subject on which he knows very little (for example, Einstein's theory of relativity or existentialism). Nonverbal behavior is tallied by the group. After these two 5-minute encounters the group members discuss the differences in nonverbal cues between the first and the second discussion. Is there more or less nonverbal behavior with the unknown subject compared to the known subject?

Nonverbal behavior with responsive and nonresponsive listener

One trainee-volunteer speaks to the other trainee-volunteer about any topic of interest to him for 5 minutes. The listener sits quietly and shows no response. The nonverbal behavior of the speaker is tallied by the group. Then the speaker discusses the same topic for another 5 minutes, but on this second occasion the listener shows interest, listens intently, and uses gestures (if he wishes) to indicate his interest. Again, the nonverbal behavior of the speaker is tallied by the group. After these two 5-minute encounters the group members discuss the differences in nonverbal cues between the first and the second discussion. Does the speaker show more or less nonverbal behavior when the listener is responsive or nonresponsive?

Nonverbal behavior with a foreigner or a child

One trainee-volunteer speaks for 5-minutes to the other trainee-volunteer who is supposed to be either a foreigner or a child. The speaker may choose either of these roles for the listener. The nonverbal behavior of the speaker is tallied by the group. Then the speaker discusses the same topic with the listener who is now within the speaker's age range and speaks the same language. The speaker's nonverbal behavior is again tallied by the group. After these two encounters the group members discuss the difference in nonverbal cues between the first and the second discussion. Does the speaker show more or less nonverbal behavior with the foreigner, the child, or with someone who speaks his language and is nearer his age?

Observation of behavior of the whole person

Role-playing. Role-playing is a method used for developing observational, recording, reporting, and interviewing/counseling skills. Role-playing involves the individual in a variety of roles and offers opportunity for self-observation as well as observation by other trainees.

The Latin origin of the word "role" is "rotulus," which refers to a rolled-up script. (The directions given in the exercises of this book may be likened to a script.) Later the French added the definition "social function" to the meaning. Putting the Latin and French meanings together, role-playing is defined as a method of performing certain acts and/or speaking certain lines according to a planned script, combined with spontaneous acts and lines. These acts or lines refer to specified social or other events. Role-playing may serve as a projective and/or therapeutic technique or as an educational or training device.

Role-playing is best accomplished if the particular skills to be learned are broken down into the specific behavior involved. For example, an individual who has been hospitalized in a mental institution for several years profits from role-playing sessions with individuals who ask him about where he has been. He practices how he will respond to people's questions about his prior whereabouts when he leaves the hospital. The job-seeker devises effective approaches for communicating with someone in a personnel office. The child becomes more comfortable when he role-plays the steps in the hospital procedure and how he should act when he goes to the hospital to have his tonsils removed. The human service worker experiments with eye contact, body position, and gestures in order to develop better listening skills.

Role-playing therefore serves several purposes:
- It provides an approach that involves the whole person in an active process of self-exploration. The person thinks, feels, and does.
- Feedback is immediate and is also a response to many different aspects of the

individual's behavior. The role-player is not only seen and heard but he has a more direct effect on others when they are in the drama with him.

- Role-playing offers a way in which an individual may learn about other people's values, feelings, and problems by experiencing events *as if he* were the other person.
- The role-player may act as a model for others by demonstrating skills, responses, problems, and so on.
- The lessons learned in the process of role-

playing one particular situation are likely to apply (generalize) to other similar situations.

Gestalt. Observing the gestalt means picking up cues about the observed person through all channels of communication and putting these all together for the pattern of behavior that emerges. Exercise 2-3 directs attention to the whole person, who in this situation is role-playing an interviewer. The measurement scale below examines the interviewer's listening and responding skills.

Measurement scale: Listening and responding skills

DIRECTIONS: Place one check next to the rating that most closely approximates your observation of the interviewer's behavior. Check under each behavioral characteristic.

EYE CONTACT: Maintains appropriate gaze, which is not a stare, nor does he look away.
1. Gaze is persistent and comfortable. _____
2. Gaze is appropriate most of the time. _____
3. Shifts gaze a little too often. _____
4. Frequently shifts gaze. _____
5. Persistently shifts gaze or stares. _____

ATTENDING: Maintains appropriate posture; bends slightly forward from waist; and maintains comfortable distance from client.
1. Persistently shifts position and moves about. _____
2. Frequently shifts position and moves about. _____
3. Shifts position a little too often. _____
4. Most of the time maintains appropriate posture. _____
5. Persistently maintains appropriate posture. _____

HAND GESTURES: Moves hands slowly and appropriately; gestures appear to be comfortable.
1. Gestures are persistently comfortable and appropriate. _____
2. Gestures are usually comfortable and appropriate. _____
3. Some signs of "jerky" hand movements, which are disturbing. _____
4. Tense, sudden movements frequently occur. _____
5. Persistently uses inappropriate and annoying hand gestures. _____

FACIAL EXPRESSION: Smiles or shows other expressions that are appropriate and pleasant.
1. Persistently maintains unpleasant and inappropriate facial expression. _____
2. Usually maintains unpleasant and inappropriate facial expression. _____
3. Occasionally has unpleasant and inappropriate facial expression. _____
4. Usually maintains pleasant and appropriate facial expression. _____
5. Persistently maintains pleasant and appropriate facial expression. _____

VOICE TONE: Voice is pleasant, sounds relaxed, and has appropriate volume for hearing.
1. Persistently has unpleasant tone that varies in loudness, either too loud or too low. _____
2. Usually has unpleasant tone that varies in loudness. _____
3. Occasionally has pleasant tone with appropriate volume. _____
4. Usually has pleasant tone with appropriate volume. _____
5. Persistently has pleasant tone with appropriate volume. _____

COMMENT: Write a brief comment about the interviewer that you think his supervisor should know.

Exercise 2-3

OBSERVING THE WHOLE PERSON

The group divides into small groups of seven trainees. Two trainees sit in the center of a semicircle so that they are visible to the other five. The five observers should have two copies of the measurement scale, which they complete at the end of the first 5 minutes of role-playing and again after the second 5 minutes of role-playing.

For 5 minutes the two trainees role-play an interviewing session in which the interviewer, a crisis teacher, is discussing an incident with the client, a student, who was sent to the "school adjustment (or crisis) room" because he called a white teacher a "honky." The discussion may take any direction the interviewer and the client wish it to take. After 5 minutes the role-players stop while the five observers rate them using the measurement scale on p. 27.

The role-players join the five other trainees and for 10 minutes they discuss the ratings and comments about the client.

The role-players resume their same roles of interviewer and client and continue the interview for 5 minutes. Then they discuss the ratings and comments about the client again and mention any degree of improvement in observations from the first to the second time.

Observation of verbal cues

The second step in learning to observe is to break down observation into a specific aspect of behavior, in this situation verbal cues. Exercise 2-4 directs attention to these verbal cues.

Exercise 2-4

OBSERVING VERBAL CUES

The group divides up into small groups of six or eight trainees. Two trainees serve as role-players. The other trainees arrange themselves into a circle with their backs to the role-players, who are in the center. Each observer needs two copies of Fig. 2-1; one he uses at the end of the first 5 minutes of role-playing and another after the second 5 minutes of role-playing.

For 5 minutes the two trainees role-play a debate in which one person takes the viewpoint that abortion should be legalized and the other person opposes the legalization of abortion. Half of the observers rate the trainee opposed to legalization and the other half rates the trainee in favor of legalization. The observers add up the tallies for each category in Fig. 2-1 and write comments about their overall impression of the verbal behavior of the role-player they have watched. The role-players join the observers and for 10 minutes the group discusses the ratings and the comments about the role-players' verbal behavior.

The role-players resume their roles and continue their debate for another 5 minutes. A second discussion of the ratings and comments is held, with a comparison of the accuracy and completeness of the first to the second debate in the observer's comments and other ratings.

Observation of nonverbal cues

Nonverbal cues are a part of behavior many people neglect, perhaps because they are so obvious. Yet nonverbal cues influence the message-receiver and often steer the conversation in subtle ways. The ability to pick up these cues should be sharpened. With practice, the trainee becomes accustomed to look for these cues. For a while this may be a heavy burden to carry. However, with experience, the pendulum swings to a pleasanter, more comfortable place in which there is an easy alertness and an understanding application of observations. Exercise 2-5 focuses on these nonverbal cues.

Exercise 2-5

OBSERVING NONVERBAL CUES

The group divides into small groups of six to eight trainees. Two trainees serve as role-players and sit in the center of a semicircle formed by the other trainees. The observers face the role-players. Each of the observers needs two copies of Fig. 2-2; one he uses at the end of the first 5 minutes of role-playing and another after the second 5 minutes of role-playing.

For 5 minutes the two trainees role-play a mother or father with her (his) son or daughter. The parent has found a bag of "white stuff" and a hypodermic syringe. The parent suspects what the white stuff may be but is uncertain. The parent confronts the child with these items and tries to discover where they were obtained. Half of the observers rate the parent and the others rate the child using Fig. 2-2. The observers also write comments about their overall impressions of the nonverbal behavior of the role-players they have watched. The role-players join the observers and for 10 minutes the group discusses the ratings and comments about the role-players' nonverbal behavior.

The role-players resume their roles and continue their probing attack and counterattack in this parent–young adult confrontation. Another 10-minute discussion is held with the addition of a comparison of the accuracy and completeness from the first to the second sessions in the observers' comments and other observations.

Interpreting pattern of behavior from verbal and nonverbal cues

Another important consideration in observations is whether *what* the person says and *how* he says it agree. In other words, does the smiling person say, "I hate you"? Does the man say, "Doesn't hurt a bit," while his eyebrows move closer and closer into a frown and his jaw and fists become tighter and tighter? Congruence, or the harmony and correspondence of verbal and nonverbal cues, is an essential point. This kind of consistency is important to observe in the client but even more important for the interviewer.

Exercise 2-6
CONGRUENCE OF VERBAL PLUS NONVERBAL CUES

For a period of 1 week, keep a notebook with you in which you write your observations of people in at least ten different situations for periods of at least 5-minutes for each observation. Jot down your ten observations and your impressions, keeping the following questions in mind:

• What about the behavior of these individuals assures you that they mean what they say?
• What about the behavior of these individuals assures you that they don't mean what they say?

Be specific in your observations and in your support of your impressions. Refer to Figs. 2-1 and 2-2 for verbal and nonverbal cues.

Bring your findings to your next group meeting to discuss with the other trainees. If you find that the other trainees are more

complete than you are in observations and supporting facts or that your findings need additional observations, you should repeat this exercise.

Comparing verbal and nonverbal behavior for cues about the observed person's honesty in his expressions brings the observational process back to the whole person. Exercise 2-7 attempts to be three things—an experimental observation, a game, and a look at the whole person.

Exercise 2-7
THE WHOLE PERSON ACTS

It would be more accurate to call this a semiexperimental observation since the only controlled variable (factor) is the input or independent variable. The independent variable that is manipulated is the choice of an island given to the trainees. The dependent variable is the resulting behavior of both the "island-chooser" and the observer, who are both trying to win their choice of an island. The game procedure is as follows:

Everyone must move to one of two islands—Etirabys or Dellortnoc. In Etirabys everyone lives in a group home (commune), sharing all possessions and doing anything he wants to do. Dellortnoc is an island where everyone lives in his own home, can have all the money he wants, and must follow Seyegib's commands. Half of the total number of trainees may move to Etirabys and the other half to Dellortnoc.

Today there is a council meeting to decide the island for each trainee. Chairs are arranged in two straight rows so that each trainee has another trainee opposite him at a comfortable distance yet close enough for conversation. Each observer-trainee needs paper and pencil; all trainees need copies of Fig. 2-3. If there are more than twelve trainees in the larger group, form one or more smaller groups. The size of the smaller group must be an even number for trainees to function in pairs.

Trainee 1 describes for no more than 3 minutes why he should go to one of the islands and should not go to the other island. Each island-chooser trainee supports his island choice.

Trainee 2 observes trainee 1, writes verbal and nonverbal cues, whether they are congruent (consistent), and also comments

	A TALLIES	B TOTAL TALLIES
1. **Verbal observations mentioned**		
2. Actual words quoted		
3. **Nonverbal observations mentioned**		
4. Different parts of body mentioned		
5. Different kinds of movement mentioned		
6. **Observation of congruence between verbal and nonverbal behavior**		
7. Supported by actual examples		
8. **Interpretations of behavior**		
9. Supported by verbal behavior		
10. Actual words quoted		
11. Supported by nonverbal behavior		
12. **Recommendations supported**		

Total score _____

Fig. 2-3. Rating the observer.

(interpretation) on the behavior of the whole person (trainee 1) in terms of his eligibility for his choice of island. Trainee 2 supports his interpretations (inferences) about trainee 1's behavior with facts gathered from the behavior he observes. It does not matter whether the observer supports or denies trainee 1's choice. It does make a difference in the observer's score if his observations or his interpretations are insufficient and unsupported.

Each trainee on side A chooses an island and supports his choice for no more than 3 minutes. Trainees on side B observe, take notes, interpret, and support their interpretations.

After all the trainees select their islands, then the observers report to the entire group. Each observer reads the verbal, nonverbal, and combined cues he has written about the island-chooser assigned to him. The observer also supports or denies the island-chooser's selection.

Trainee 1 places a mark (tallies) in column A each time observer 2 mentions a particular item. The total number of tallies next to each item is written in column B. The total of all the scores in column B is written at the bottom of column B. This is the observer's score. A sample rating is shown in Fig. 2-4. The other island-choosers rate their assigned observers in the same way as trainee 1, as shown below:

Island-chooser		Observer
1	rates	2
3	rates	4
5	rates	6
7	rates	8
9	rates	10
11	rates	12

Trainee 1, or any of the other island-choosers rating the observers, may challenge the observer's comments. The observer must support his comments with his observations to the group's satisfaction. The observer's and island-chooser's keen observation is the crucial element in accurate scores.

If there are twelve trainees, six of them may choose their islands. This means three island-choosers and their three observers who have the highest scores.

Mechanical devices

Observation over the telephone must concentrate on verbal cues. The tape recorder,

	A	B
	TALLIES	**TOTAL TALLIES**
1. **Verbal observations mentioned**	~~////~~	5
2. Actual words quoted	//	2
3. **Nonverbal observations mentioned**	///	3
4. Different parts of body mentioned	///	3
5. Different kinds of movement mentioned	//	2
6. **Observation of congruence between verbal and nonverbal behavior**	/	1
7. Supported by actual examples	0	0
8. **Interpretations of behavior**	~~////~~	5
9. Supported by verbal behavior	//	2
10. Actual words quoted	/	1
11. Supported by nonverbal behavior	//	2
12. **Recommendations supported**	0	0
	Total score	26

Fig. 2-4. Sample observer rating.

however, may focus on verbal cues only (the audio tape recorder), visual cues only (videotape with the sound turned off), or on a combination of visual and auditory cues (videotape with both sound and picture). The following exercise involves observation of verbal cues only.

Exercise 2-8
TELEPHONE OBSERVATION

Two trainees volunteer as role-players, one of whom is the client, the other the interviewer. The remainder of the group (the observers) arrange themselves so that half are seated in a row on one side of the room and half on the other side of the room with the role-players in between. All observers turn their chairs around so that they cannot see the role-players.

Procedure A

Role-players arrange their chairs back to back so they can hear, but not see, each other. They select one of the following situations and assume appropriate roles.

1

Tom is 16 years old and rather uncertain of how he stands with Eva. He wants to ask her to his junior prom. He calls her to invite her. Eva may accept or refuse Tom's invitation.

2

Sandra is calling her home to tell her parents that she has been "busted" for "possession." She wants to ask her mother to arrange bail, but her father gets on the phone.

3

Mrs. T. is calling her lawyer to ask him to do something about her husband who has just strap-whipped their 6-year-old son, Terry. This is the fourth time in 2 months and Terry has red welts from the strap. Mrs. T. asks the lawyer how she can get her husband to leave the house.

• • •

Observers, using the list of cues in Fig. 2-1 as a reference, record as many verbal cues (of Tom, Sandra, or Mrs. T.) as they can during the role-players' conversation.

Procedure B

After 5 minutes of the telephone conversation, the entire group gets together to examine the accuracy and completeness of observations and to arrive at some interpretations of the observed behavior. This discussion lasts 10 minutes. Then the same role-players continue their conversations, but the observers change the focus of their observation from Tom, Sandra, or Mrs. T. to Eva, the parent, or the lawyer. This second telephone conversation lasts 5 minutes and then the group once more gets together for 10 minutes to discuss the observations and to arrive at some interpretations of the behavior.

The tape recorder is a different technological (mechanical, electronic) device than the telephone since it provides both feedback and serves as a recording tool. Audiotaping is proposed for the rest of the exercises.

RECORDING

Alertness to the many cues around the observed person and the observer is one part of observation. However, if this information (data) is not recorded, chances for errors in remembering are increased. The longer the observer waits to write his notes, the greater the possibility for incorrect data. Recording goes beyond the written word. The efficient observation indicates the relationships involved in the sequence of the "happening" before, during, and after the behavior.

ABC sequence of recording

The antecedent event (A), behavior (B), and the consequent event (C) represent the ABC sequence of recording behavior. These three events are intertwined and often occur in rapid succession, almost simultaneously (within seconds of each other).

Antecedent event

Recording in terms of ABC provides the observer with points of reference from which he may determine the events that encouraged certain behavior to begin and the events that reinforced the behavior to continue. Thus the observer may be more helpful in pointing out to the teacher the cues from his behavior that prompt Kay's

constant motion to linger on—up and down Kay moves, in and out of her seat, when the subject she hates must be worked at "quietly in her seat." Kay's behavior (leaving her seat) reveals the reinforcements (rewards) she obtains from her behavior. First, she avoids the necessity to "play around" with sums, subtractions, and divisions (self-reinforcement). Second, the teacher pays attention to Kay (teacher's reinforcement). At no other time during the day does Kay become the center of attention. Stating the ABC of behavior, therefore, helps explain the antecedent, (beginning event), the resulting behavior, and the consequent reinforcement (ending of the event).

Part of the antecedent, the behavior, and the consequence is overt; that is, it can be seen or heard. Part of these three aspects is not observable; it is covert and cannot be directly seen or heard. Since overt factors are observable, they are objective and reportable. Covert factors are more or less hidden and must be guessed at or inferred from the observable behavior.

Often it is simpler to notice the consequent event rather than the event that precedes an individual's behavior. One reason for the difficulty in isolating the antecedent event is the tendency to attempt to establish cause-and-effect relationships. However, behavior does not result from one isolated stimulus (antecedent event) but instead from a number of stimuli (events), which are interrelated into a stimulus-situation (pattern of antecedent events). Another name for an antecedent event is an independent variable. In an experimental situation occurring in the laboratory the independent variable can be manipulated (changed) so that the changes in behavior, the dependent variable, may be observed. In real life it is much more difficult, if not impossible, to control factors so that only one item (variable) changes. Exercise 2-9 begins this more objective recording process with consideration of the antecedent event and the resulting behavior.

Exercise 2-9
ANTECEDENT EVENT
AND BEHAVIOR

Read the three examples of behavior that follow and fill in the accompanying chart

	ANTECEDENT EVENT	BEHAVIOR
Cindy		
Timmy		
Bud		

to show the antecedent event and the behavior.

1

Cindy is a "sitter" who gives a vacant lot as her address. She is arrested during her walk along her "turf" (assigned territory by the man who keeps her, her "pimp") after she sells a key to her apartment with a promise of sex relations to Max, an undercover policeman. When Max roughly grabs her, she scratches his face on the left side with four of her long fingernails.

2

Timmy's mother stands over him at the breakfast table playing the "transportation game." She says "whoosh" as she plunges the spoon ("airplane") into the oatmeal, carries it over Timmy's head into the air, and lands it in the "hangar" of Timmy's mouth. After each "airplane" feeding, she picks up the glass of milk and holds it to Timmy's lips. Some of the milk flows down his throat. Some of the milk splatters down his chin onto his striped polo shirt. Timmy shakes his head from side to side but his mother continues feeding this 4-year-old whenever his mouth reaches her outstretched hand. Timmy begins to slowly release some of the oatmeal from his mouth; it slithers down his mother's hand as she tries to "get the airplane into the hangar." "Timmy, you are a bad boy!" his mother screams as she quickly removes her hand and stops feeding Timmy.

3

Bud is explaining to the judge the reason for his forging three checks, "It's like this, judge—your honor. There is this woman, calls herself a witch. She mixes a strange

purple brew—don't remember anything after you drink it. She made me drink it. Then I had to do anything she asked me to do—don't remember what."

• • •

After completing the chart, discuss what you wrote with other members of the group.

Behavior

The antecedent event answers such questions as, "What happened before the activity (behavior)?" "What was the beginning of the event?" "What stimulated the activity?" "What spurred the individual to action?" and "Who or what began the activity (behavior)?"

Some other points to remember about the antecedent event (the stimulus, the excitant, the independent variable) is that it may be obvious—something or someone outside that can be seen or heard. Therefore it is overt. On the other hand, the spur to activity may not be seen or heard (covert), such as an inner excitant or impulse (a feeling).

Behavior is what an individual does or says after the antecedent event. This observed behavior should be stated in such a way that it describes the action seen or heard. Exercise 2-10 focuses on the difference between an observable and measurable statement about behavior and a vague description.

Exercise 2-10
BEHAVIOR—OBSERVABLE AND MEASURABLE

Some of the brief statements about behavior in the chart on p. 34 are specific, obserable, and measurable. Others are vague and nonspecific. Place an "x" in the "specific" column of those statements you con-

STATEMENT	SPECIFIC	NON-SPECIFIC	CHANGE TO SPECIFIC
Example: He gives me a pen.	x		
Example: He is nice to me.		x	
1. She puts away her blocks.			
2. He enjoys himself.			
3. You will know.			
4. You understand.			
5. They all drink to Ted.			
6. He changes his mind.			
7. She changes her clothes.			
8. He sets the table.			
9. He feels good.			
10. She screams at him.			

sider observable and measurable and in the "nonspecific" column for those you think are nonobservable and nonmeasurable. In the third column, change the nonspecific statements to a specific statement. Note the two examples.

After completing the chart, discuss your opinions with the other trainees in the group.

Another detail relating to behavior is the setting in which the behavior occurs. The setting shares in the effect of the antecedent event and touches upon such questions as:
- What is going on (the event or events) around the observed individual (external factors, physical setup)?
- Who are the people around the observed individual?
- What are the demands and expectations of people and things around the observed person?

In order to demonstrate how the setting makes a difference, the behavior and antecedent events in Exercise 2-9 are used and the setting is to be supplied by each trainee.

Exercise 2-11
THE SETTING OF BEHAVIOR

Read the situations in the three examples in Exercise 2-9 and devise settings that alter either where the observed person is or who is around him at the time of the antecedent event. Fill in the chart on p. 35 with short phrases showing how the behavior is the same or is altered by the setting. In the fourth column, explain what makes the change in behavior.

After you complete your chart, discuss your ideas with other members of the group.

Consequent event

Consequences are the events that immediately follow behavior. Often individuals who set out to change behavior actually reinforce the behavior they want to change. For instance, the mother says to her child, "Again . . . you didn't come in when I called you. Wait until your father comes home. He'll get you!" The threat is there and Mother may tell Father but the consequence has little or none of the planned effect that the mother hopes to accomplish. The consequence is too far removed from the behavior. The influential consequence is the mother's admission that she is not doing anything about the undesired behavior. In this way, Mother is strengthening (positively reinforcing) the undesired behavior. Behavior continues mainly because of the consequences (effects) that it produces. Thus observation of the relation between behavior and its consequences (contingency) becomes very important for the human service worker. Exercise 2-12 examines contingencies.

	SETTING	ANTECEDENT EVENT	BEHAVIOR	EXPLAIN
Cindy				
Timmy				
Bud				

Exercise 2-12
CONTINGENCIES—BEHAVIOR PLUS CONSEQUENCE

In the example of behavior below, insert the antecedent and the consequent events. Write one consequent event that is likely to continue the behavior and one that would tend to decrease and/or stop the behavior.

Differentiating between pure observation and pure inference

Greater accuracy is achieved when the observer is able to distinguish among three kinds of note-taking.

Pure observation

Pure observation gets the facts, the data, from looking and listening. Such data states, "John lost his job," "Marcia raised her hand," and "The woman coughed." These are the kinds of data that can be verified and about which observers from similar cultures would agree. However, the cold, unembellished, unfeeling facts about human behavior leave out the inferences or the feelings that color the observations and make them more personalized.

Pure inferences

Observations that are totally based on impressions, hunches, and hypothesis tend to lose something because of their subjectivity. Inferences arise from the observer's reactions—the observer's feelings that accompany the objective events. Too many inferences may color the facts with the observer's bias and often distort what the observed person is actually saying or doing.

A combination of inferences and observations

The most effective notes about observations have inferences supported with objec-

ANTECEDENT EVENT	BEHAVIOR	CONSEQUENT EVENT
_____ _____ _____ _____	Marie is 10 years old but has a mental age of 5. She grunts or points when she wants something. (The goal is to encourage her to speak.)	_____ _____ _____ _____

tive evidence from observations. The examples indicate the difference between pure observation, pure inference, and a combination of the two.

Example 2-1
COMPARING PURE OBSERVATION AND PURE INFERENCE
Pure observation

Mary hits John three times. Her lips are in a thin line and her eyes stare. Mr. Jones says, "I saw Mary angrily hit John." (A recorded comment made by an individual other than the observer is a secondary observation.)

The teacher told Tommy to open his book and to start reading. Tommy lowered his head and did not open his book. The teacher repeated her request in a louder voice. Tommy, head still lowered, after 1 minute, opened his book and began to read slowly.

Pure inference

Mary angrily hits John.

Observer writes: "Mr. Jones may have told her about Mary's anger at John."

Tommy is obviously hostile toward the teacher. He refuses to do what she asks him to do. The teacher doesn't realize that she is encouraging Tommy to be more hostile by her own hostility. Although Tommy finally does what the teacher commands, he remains hostile.

Combination observation and inference

Mary hits John three times. Her lips are tightly held in a thin line and her staring eyes glare her anger.

Mr. Jones says, "I saw Mary angrily hit John." Mr. Jones' opinion is supported by other observations. Mary's "tight" lips, glaring eyes, and hitting John show her anger.

The teacher told Tommy to open his book and start reading. Tommy lowers his head and does not open his book. (Tommy's apparent refusal to do what the teacher asks him to do suggests hostility.) The teacher repeats her request in a louder voice. (Teacher's hostility and attention to Tommy's inappropriate [undesired] behavior are probably reinforcing Tommy's behavior.) Tommy, head still lowered, after 1 minute, opens his book and begins to

read slowly. (Tommy gave in, but his feelings of hostility appear to continue. Tommy's relationship with the teacher needs further exploration.)

REPORTING

Two steps must precede any reports about behavior—observation and recording. Reports extend beyond these steps to the analysis of behavior.

Multiple hunches or hypotheses to explain behavior

Hunches—stated in the plural—make certain assumptions about human behavior. These assumptions are the following:
- The causes of behavior are multiple and interrelated.
- The same behavior may have different causes.
- The same cause may result in different behavior.

The causes of behavior are part of a system of causes. For example, an explanation of Tommy's behavior described in Example 2-1 is oversimplified when the teacher is made the "fall guy." Tommy, the "hostile" boy, may be tired, hungry, upset about his mother or father, or thinking about Bud who called him a "nigger." As the observer considers more explanations or hunches about Tommy's behavior, he is becoming aware of the multiple and interrelated causes of behavior.

Mary may be hitting John because she is angry at someone else. John just happened to be the switch that turned her anger into action. On the other hand, John may have reinforced Mary's angry display because he enjoys it just as much as she does. This is a game called "anger," which they often play as part of their boy-girl relationship. Furthermore, this kind of interpersonal transaction—anger—is the only way that Mary has learned to show her interest in boys.

Support from observations for hunches

Accordingly, in order to analyze behavior, multiple (many) hypotheses must be considered to explain behavior. The next step is an examination of observations to discover which of the hunches are supported. These hunches or inferences that arise from the observer's interpretations of the feelings and reasons for the observed person's behavior must be handled cautiously. An example of a twisted tale arising from unsupported hunches emerges from the possible interpretations even with a simple, direct statement such as, "Marcia raised her hand." One observer says that Marcia raised her hand to strike Hank. A second observer is certain that Marcia raised her hand because her deodorant stung and she wanted to air her underarm. A third observer believes that Marcia sees Alan in the distance and is getting ready to wave to him. An illustration of how all of these hunches are supportable comes from Marcia's version of what happened. She tells this to the fourth observer who does not stop with his own hunch but attempts to get more facts.

1

MARCIA: Oh, my hand. *(She smiles.)* That really was far out. I was using this new de-odor jazz . . . supposed to not sting,

ye know . . . hah . . . stings . . . wow! Was I glad that Hank came along . . . asked him whether the de-odor really worked . . . at least *that* . . . raised my hand . . . felt sorta funny about it . . . saw Alan and waved to him . . . .

2

MARCIA: Oh, my hand. *(She smiles.)* That's a new kind of salute we've gotten together . . . means we're on the scene . . . .

Observations that do not support hypotheses

Marcia's reasons prove that caution in conclusions is the key word for any analysis of behavior. In order to arrive at even tentative conclusions about feelings and reasons, the observer must not only try to support his hunches from his written notes but must also try to determine whether

Specific behavioral event: _____

MULTIPLE HYPOTHESES (interpretation/inference about observed person's behavior)	SUPPORT (observation/facts supporting interpretation)	REFUTE (observation/facts not supporting interpretation)
1. _____ _____ _____	a. _____ _____ b. _____ _____ c. _____ _____ (Add more items if available.)	a. _____ _____ b. _____ _____ c. _____ _____
2. _____ _____ _____	a. _____ _____ b. _____ _____ c. _____ _____ (Add more items if available.)	a. _____ _____ b. _____ _____ c. _____ _____
3. _____ _____ _____	a. _____ _____ b. _____ _____ c. _____ _____	a. _____ _____ b. _____ _____ c. _____ _____

Example 2-2
SUPPORTING MULTIPLE HYPOTHESES: SPECIFIC BEHAVIORAL EVENT
Specific behavioral event: Joe lost his job.

MULTIPLE HYPOTHESES	SUPPORT	REFUTE
1. He is depressed.	a. He says, "I feel lousy." b. He begins to cry. c. He bangs his hand on the desk in front of him. d. He looks straight ahead. e. His shoulders slump and his back becomes rounder. f. He goes to the bar across the street.	a. He smiles broadly as he says "good-bye" to the people in his office. b. He tells jokes to the men at the bar across the street.
2. He is happy.	a. He smiles broadly as he says "good-bye" to the people in his office. b. He goes to the bar across the street and tells jokes to the men.	a. He says, "I feel lousy." b. He begins to cry.
3. He is angry.	a. He bangs his hand on the desk in front of him. b. He looks straight ahead. c. He says, "I feel lousy." d. He goes to the bar across the street.	a. He begins to cry. b. He smiles broadly as he says "good-bye" to the people in the office. c. He tells jokes to the men at the bar. d. His shoulders slump and his back becomes rounder.

there are any observations that can punch holes in the hypotheses.

The chart on p. 37 that is based on the approach to analysis of observations by Prescott (1957) is useful for practicing analysis of observations of field experiences (practicum) or for notes taken about incidents in the training group or other places. The important aspect is that there must be ample opportunity to observe in order to support or deny the hunches.

Use Examples 2-2 and 2-3 as samples of the procedures that may be used to determine possible explanations (hypotheses) for a specific behavioral event or repeated behavior (recurring).

Continue with your hunches. More hunches usually mean you are getting deeper into the possible explanations.

Multiple hypotheses may be constructed for one particular behavioral event, such as Joe loses his job, Marcia raises her hand, or Susannah shuts her mouth when the spoon reaches her lips. Hunches or hypotheses may also be formulated for behavior that is repeated, *recurring behavior*. There

is one additional procedure for the analysis of recurring behavior. Observational notes must be sifted to discover the behavior that is repeated several times. Example 2-3 examines Tommy's behavior in the classroom (see p. 39) in order to discover his feelings and possible explanations for his behavior.

What other hypotheses might there be for Tommy's behavior? How may these hypotheses be further verified or discarded? Which of the hypotheses seems to have the most support? Which of the hypotheses seems to have the least support?

Tentative implications of findings from hunches that are supported and those not supported

Constructing multiple hypotheses and then supporting or discarding these hypotheses from the facts gathered during observations leads to tentative conclusions. The observer pulls together those hypotheses that appear to have the most observational support. Then the observer decides what other information he needs and what other

Example 2-3
SUPPORTING MULTIPLE HYPOTHESES: RECURRING BEHAVIOR

MULTIPLE HYPOTHESES	SUPPORT	REFUTE
1. Tommy feels hostile toward his teacher.	a. Tommy does not open his book when the teacher asks him to do so. b. Tommy does not look at teacher when she talks. c. Tommy reads slowly with his head bent.	a. Tommy complains of headaches on three observations. b. Tommy rubs his left eye on four observations. c. Tommy walks up to the blackboard to "see better."
2. Tommy feels uncomfortable when he reads.	a. Tommy moves about in his chair during reading time. b. Tommy says, "I can't" when asked to read.	
3. Tommy does not know how to read.		a. Tommy reads most words, particularly when he gets close to the book.
4. Tommy does not see well. He needs corrective lenses.	a. Tommy complains of headaches. b. Tommy comes up to the blackboard to see better. c. Tommy rubs his left eye. d. Tommy reads most words, particularly when he gets close to the book.	
5. Tommy has perceptual difficulties.	a. Tommy reads a "d" for a "b." b. Tommy reads "was" for "saw." c. Tommy rubs his eye and complains of headaches.	

observations he must make before he can consider his interpretations of behavior correct, or at least tentatively correct. Slowly, but surely, the observer proceeds through the scientific method for analyzing behavior.

This does not suggest that observation is the beginning and the ending role of the human service worker. Instead, the ability to observe accurately, to analyze and arrive at interpretations cautiously, are the foundation for helping. With practice, the trainee embraces the ability to observe and hypothesize. He learns how to accept or reject these hypotheses. Meanwhile, he is functioning as a helper to the client while with the client and analyzing his observations later so that he is better prepared to help the client and is not just walking on air.

Summary of Part One

The first chapter in Part 1 explores the centipede-like services of the human service worker. A distinction is made between the human service worker who may be anyone who gives attention and assistance to another person and the professional human service worker who is equipped with knowledge, skills, and a code of ethics. Another distinction is made between the generalist, who concentrates on the needs of the whole person in his environment, and the specialist, who narrows his care-giving to only certain needs. Careers for human service workers should allow for growth rather than become blind-alley jobs. In order to accomplish career growth, there must be a united effort by educators and employers for lifelong updating and upgrading of knowledge and skills. This opportunity for lifelong learning encourages both career growth and the fulfillment of individual potentials. The professional self is an outgrowth of training, self-understanding, and a code of ethics that respects the rights of the person being helped.

The second chapter presents the beginning layers of the helping process—observation, recording, and reporting. Part of the self-understanding the trainee must develop emerges from his definition of his learning style and of his identity-concept. The second part of this understanding is his increased alertness to the factors that may influence the accuracy of the trainee's observations. Two broad categories of influential factors are mentioned, the physical and psychoecological areas. These take into account the sensory acuity and the health of the observer as well as the many other environmental variables.

General observational procedures are related to the degree of involvement of the observer with the observed person and the type of observation used. In terms of involvement, observations may be gathered from secondary sources, spectator observation, participant observation, or the proximity required in the interview/counseling psychotherapy relationship. The types of observation are categorized as casual, naturalistic, standardized, specific goal, experimental, or some combination of these varieties. The development of observational skills requires alertness to the verbal and nonverbal cues as well as to the behavior patterns that results when verbal and nonverbal behavior are combined.

Observations may be recorded by means of written notes or on tapes. Tape recordings serve as both a means of note-taking and as a training device for feedback. The accurate observer records his observations according to the ABC of behavior. This includes the antecedent events, the behavior, and the consequent events. In addition, inferences are supported by observations to make certain that the observer's interpretations of behavior are supported by what he saw and heard. Reporting starts with observation and recording and gets into analysis. Analysis of behavior depends on knowledge of human behavior in general and more specifically on exploring multiple hunches (hypotheses) that may explain the behavior. The record of observations must be searched for behavior that supports and that does not support the hunches. This method provides a more scientific approach to the study and understanding of behavior.

PART TWO

THE WHAT AND HOW OF INTERVIEWING

HOW TO GET OUT OF THE CLIENT'S WAY

The salesperson says it to the customers examining the double-knit suits hanging on the "Sale Today" rack:

"May I help you?"

The receptionist in an employment agency office says it to the puzzled job-seeker:

"May I help you?"

The Frenchman says it to the weary sightseer wandering through an unknown street, confused by a foreign language:

"May I help you?"

• • •

To help has many meanings:

- For the salesperson it may signify an added commission and making points with the supervisor.
- For the receptionist it serves to direct to the personnel interviewer a body that may be cluttering up the office.
- For the Frenchman it may be the selling point for other tourists to visit the "friendly" city.

But for the effective interviewer it means *caring.*

CARING is the characteristic to be sought in the interviewer. Care means giving serious attention to someone. The cared-for person feels that the caring person respects and accepts him as worthy. The interviewer who cares encourages the other person to explore his beliefs and the influences that are helping or hindering him in making a decision or in solving his problems. The interviewer who cares is genuine, warm, and understanding and avoids intruding his ways of living (life-style) on the psychological field of the person being interviewed.

Two essential ideas spring forth from these ideas about caring. First, it is obvious that helping *may* include caring, but caring *always* includes helping. Second, to accomplish being with and not acting for another person is not easy.

CHAPTER THREE

TYPES OF INTERVIEW RELATIONSHIPS

To help has many meanings. (Photo by Jerry Sloan, City of Chicago, Department of Human Resources.)

ACTION/REACTION

In an action/reaction interview, one individual acts upon another—asks a question, for example. A reaction in the form of an answer is not necessarily desired, expected, nor even awaited. A nonrelationship exists.

Think of how many times during the day the question "How're you doin'?" is flung through the emptiness between people. Sometimes an answer is flung back, "Fine" or "Okay," or perhaps just a mumble, a smile, or silence is the answer. An empty question is met by an empty response. No one really cares; they are just creatures of habit performing a social duty.

INTERACTION
Cross-communication

An interaction interview allows for more regard of people as people rather than things. This kind of interview appreciates the possible effect of each person in the interpersonal situation. Interaction prompts cross-communication (back-and-forth, two-directional) and cross-influence.

Psychodynamic forces

Although the interaction type of interview is a step ahead of the action/reaction interview, one ingredient for an effective interview is still lacking. This ingredient stems from the ever-changing psychodynamic (interpersonal) forces within the interview that may make the interview a hotbed of emotions.

These psychodynamic forces are the ideas, impulses, and emotions that develop, influence, and change each participant in the interview; these psychodynamic forces push/pull the client and the interviewer because they are both humans with particular ways of viewing the world and of viewing one another (frames of reference).

Measuring interactions

If these psychodynamic forces are disregarded, one way to determine the proficiency of the interviewer is to measure the frequency of verbal and gestural interactions within a specified time period. By counting the number of times the interviewer speaks and/or the number of words spoken as well as the number of interruptions, the degree of client-initiated and interviewer-initiated verbal behavior may be determined. The same numerical determination may be conducted for gestural or other nonverbal responses of the interviewer.

Fewer verbal and nonverbal expressions by the client would indicate less client participation and probably greater interviewer interference. However, the essential elements that are glossed over by this counting method may be those that would decide the outcome of the interview. Example 3-1 provides some forceful concerns that may influence the dynamics of the interview.

Example 3-1
THE INTERACTION INTERVIEW: SOME CONCERNS

The interviewer, Mr. N., has been requested to speak to Marty, a 12-year-old boy with a mental age of 6. Marty's housemother reported that Marty is frightened by something. Marty's record reveals that he is able to speak, though not too clearly, and that until now he has been able to take care of his toileting and feeding needs.

INTERVIEWER (MR. N.): *(As Marty enters the room.)* Hello, Marty, remember me? I'm Mr. N.

MARTY: *(Mumbles something while keeping his head down.)*

MR. N.: Say, Marty, here's a chair for you. I'll put the chair right here so we can see each other. *(Mr. N. moves his chair to face Marty.)*

MARTY: *(Goes over to the chair and leans against it.)*

MR. N.: Marty, I have a candy someplace in my drawer. *(Searches in desk drawer and, finding the candy, holds it out toward Marty.)*

Marty does not move toward the candy. Without a word, Mr. N. takes Marty's hand and walks with him to the housemother, who is waiting outside the office. Mr. N. shrugs his shoulders and puts his hand up, palm upward, in a gesture the housemother interprets as frustration.

Many questions arise when this interview is examined.

- How does Mr. N. feel about the mentally retarded individual?
- How does he feel about Marty in particular?
- Was everything done to make the atmosphere of the interview comforting and secure for Marty?
- If the verbal and nonverbal expressions of

the interviewer were counted, would an adequate evaluation of the effectiveness of the interview be obtained?

- Are there other facets of the interview that should be scrutinized?

Blocked communication and insensitivity to Marty interfered with a favorable interview atmosphere. Offering (bribing?) Marty candy was not enough. Politely introducing himself to Marty was not enough. Neither of these attempts takes into account the covert (hidden) contents of the interview situation.

Even though Mr. N. went through the motions of trying to relate to Marty, he failed. He may have failed because he was not really sufficiently interested in Marty as a person or his failure may have been due to his preconceived notions of what might happen during the interview. Part of his failure may have been due to the quick acceptance of failure—giving up too soon before Marty had a chance to respond.

Separation of interactors

It is even more important to realize that interviewing as an interaction implies that there are two separate and independently existing persons who may be sharing the same moment in space but not the same world.

TRANSACTION
Constancy of change in social system

The interview as a transaction may be drawn as a circle to symbolize the constancy of change that continues throughout the interview. A circle also represents unity and the wholeness of the living processes going on within the interview. These living processes, feeling and behaving, become an interpersonal unity—a social system in which the interviewer and the client establish rules, games, barriers, or openness in the drama that unfolds between and around them. The interview as a transaction becomes a laboratory in human relations, during which both the client and the interviewer experience self-discovery and personal growth.

Factors influencing interview

The structure and functioning of the interview as transaction are influenced by such factors as the following:

1. Status. Refers to culturally and individually determined positions that stem from sex, age, race, and education. The status of the person determines the rights and obligations of the person. How the status of the interviewer and client is handled may help or hinder the interview process.

2. Roles. Refers to culturally and individually determined patterns of expected behavior that may help or hinder an interview, depending on the meaning and worth of the roles to the individuals.

3. Interrelatedness and interrelating. Interrelating and interrelatedness are distinguishable. Interrelating occurs when the participants in an interview consider themselves *separate individuals* interacting with one another (an interaction interview). Interrelatedness refers to two individuals working toward a *mutual relationship*. Interrelating assumes there is respect, worth, and perhaps dignity for the client. Interrelatedness goes farther by adding equality to respect, worth, and dignity.

Status, roles, and interrelatedness are all part of the social system of the transaction. In this society of the interview the interviewer and the client weave an ever-changing pattern of responses.

Partners in interview pattern

This concept of transaction, of the interview as a social system, includes interaction but goes beyond it. The client is welcomed as an active partner in the interview. The client is accepted as a person with knowledge, attitudes, values, and potentials. The client is considered to be someone for whom the awakening and fulfillment of creative potentials are more important than passive adjustment.

The *here and now* becomes the center of the pattern evolving from the interview transaction. The client's present conflicts, orientation toward goals, and fulfillment or nonfulfillment of goals are not fitted into any theory in order to explain what is happening. Such a straitjacket makes a poor fit because of the uniqueness of each client. Uniqueness signifies that no two individuals are alike. There may be some general similarities among people from similar cultures, yet each individual's life experiences plus physical factors induce each person to see, hear, and feel, a little, or even a great deal, differently from everyone else, even in the same culture. Thus uniqueness is linked to perception.

Perception

Perception is the key to understanding what goes on in an interview. The hippie may see the policeman as a pig harrassing him for doing his thing. The citizen espousing law and order may see the policeman as a person who wields a strong arm in keeping those "dirty longhairs" in line. The medically oriented human service worker may think "mental illness" is the best phrase to describe a troubled person. The learning/behaviorally oriented worker may believe "mental illness" is an undesirable phrase for an individual's inappropriate habit patterns.

The way in which an individual sees the world, what has happened, and what will happen to him has a long history. Although perception of the world originally comes through the senses, these sensory impressions are modified and given meaning by the learning that takes place during life experiences. In other words, perception (seeing, hearing, smelling, tasting, and touching) is not the same for everyone. It is each individual's associations with these impressions that make them different.

To start at what may be the beginning, observe what happens to the infant. For the infant the mother's breast or the bottle provides warmth and comfort. These pleasant feelings arise from the satisfaction given by milk flowing into the baby's mouth and to its stomach. Later the breast or the bottle, and even later the mother herself, takes on meaning over and beyond the nourishment provided. The meanings of trust/mistrust, of good/bad, and of worthy/unworthy become associated with the mother. The word "mother" itself takes on these meanings as well as the emotional connotations originally associated with the person. Some of these meanings and emotional overtones affect the way the infant feels about himself and later about others.

Self-fulfilling prophecy

From the beginning the growing infant constructs a self-fulfilling prophecy. Mothers in general, a particular woman, or perhaps all women are presumed to be good/bad, interested/disinterested, and dependable/undependable according to the perceptions

that have sprung from these early experiences. It is from these early learnings and their later modifications that self-fulfilling prophecies are built.

Self-fulfilling prophecies unfold from the world each individual creates from experiences in dealing with his environment. The individual's reality and resulting behavior tend to be shaped by the way he views the world and, more pertinently, the interview situation. In turn, individuals act (behave) as if the world is as they perceive it to be. Even more important, the individual selectively, and perhaps mistakenly, appraises what happens around and to him in terms of this individualized idea of reality.

Prejudices

The appraisals of people and things are brought into the interview room and affect the behavior of both client and interviewer in positive and negative ways. Prejudices, both for and against people, become the "eyes" through which each individual perceives the other. These prejudiced eyes spur the self-fulfilling prophecy to come true.

The loner who believes himself incapable of establishing a relationship with someone will act as if these prejudices were facts and will seek responses in others that support his point The prophecy becomes true—not because it *is* true, but because the individual has pushed the prophecy into becoming real.

The parent who assures and reassures the son of his brilliance by repeating over and over again, "You are so capable. Why don't you try harder?" may be encouraging the son to failure. The son has two people to prove that he is not capable—himself and his parent. In this kind of failure he knows he can succeed. Hasn't it been proved over and over again?

The person who believes that "all people on welfare want to do is to live off government money" will find those characteristics in the "lazy" behavior of the sloppily clothed slum-dweller. This person will prove he is right. "Why, I even offered the parasite a job, and the lazy bum just said he had no shoes to wear to the job. What a feeble excuse."

The dangers of these prejudices that are supported by self-fulfilling prophecies become even more serious to the interviewer who cannot see the client without the jaundiced eye of prejudgment. Awareness of these prejudices sometimes helps. Exercise 3-1 is directed to increase such awareness.

Exercise 3-1
WORDS AND FEELINGS

On the top of a piece of paper write the word "race." As rapidly as you can, write the words that come to mind when you think of "race." Take no more than 2 minutes to do this. Put this paper aside. Now take the word "equality." Follow the same procedure and then put this paper aside. Take a third sheet of paper and follow the same procedure with the word "ghetto." On the fourth sheet of paper write the word "power" and proceed as for the previous three words.

Now compare the four sheets. Make a list of what you consider to be the favorable (positive) and the unfavorable (negative) associations you have made with each word. Then answer the following questions.

1. Do you note any similarities or differences in your four lists?
2. Are there any differences in the numbers of positive and negative words and in the kind of associations listed for each of the four words?
3. How would you explain your lists of words?

Discuss and compare your list with the lists of others who have done the same exercise; this is often an eye-opener since you see different viewpoints. You may also try this exercise with other groups of words such as "war," "camp," "draft," and "foreign."

Reciprocal effects

Since both the client and the interviewer are discovering each other, mutual progressions and regressions (reciprocal effects) are inevitable. Improvement or deterioration of the interview situation depends on the level of functioning of the client and particularly of the interviewer as well as on the ways in which they perceive (see and hear) each other.

There are rules and experimental findings about behavior, norms, standards, normality, and abnormality. In the interview as a transaction these standards and principles based on observation of many people are not rigidly applied to a particular client but are instead used as guides. These guides parent ego state, adult ego state, and child ego state. The client becomes more aware of the methods of manipulation within social situations. These methods are demonstrated in the way the client and others communicate (verbally and nonverbally). Thus transactional analysis of the games clients play (Berne, 1964) to defend themselves or to attain their ends (self-gratification) reveals more about their transactions and points out the lengthy and complex operations or maneuvers the client engages in with others. The client outlines his life-plan script (psychological drama) on which he bases his maneuvers. Script analysis helps the client understand the elaborate fantasies that may be interfering with a freer and more open existence. The script also reveals the private logic from which the client's behavior flows.

Transactional analysis clarifies the interplay among people in daily living as well as within the interview situation. The changing spectrum of relationships, the kaleidoscope of changing patterns within the interview transaction, may be partially explained by the degree of dominance (parent), dependence (child), or empathic understanding and genuineness (adult) expressed by the client and the interviewer.

Games may be played within the interview to cover up certain anxieties as well as to open or to conceal needs, requests, or demands of the client or the interviewer. Games also reflect the effect of self-fulfilling prophecies discussed earlier in this chapter.

Since the interview is a "mini" social event, it cannot be likened to a color snapshot of a frozen moment. The only constancy in a social event is change. Change is always occurring; therefore change can be depended on to be constant. This is a significant idea, for it is this idea of change that provides a positive, forward-looking approach to interviewing and, in fact, to all interpersonal relationships.

A conviction that what *is* will continue in the same fashion makes for the static view of behavior expressed in Examples 3-2 and 3-3. In both examples there is evidence of bias, self-fulfilling prophecies, and interpersonal games of defense and gratification. Closer examination may also reveal the attitudes conveyed by the parent, adult, and child ego states.

Example 3-2
THE SWITCHBLADE:
WEAPON OR STATUS SYMBOL?

Al and Johnny were talking as they climbed the steps to their classroom. Al removed his switchblade from his coat pocket and demonstrated the speed with which the button released the blade. Mary, walking nearby, heard the snap of the knife and, turning swiftly, saw the blade as it sprang out. Al and Johnny continued to talk and walk. Mary stopped, looked around, and pressed close to the railing.

Exercise 3-2
THE SWITCHBLADE:
WEAPON OR STATUS SYMBOL?

Four trainees volunteer to role-play. Al, Johnny, and Mary are sent to the school counselor. Role-play an interview between the counselor and the three students to demonstrate three different ways of handling the situation from Mary's perception of the switchblade as a weapon of attack, from Johnny's perception of the switchblade as a symbol of a cool cat, and from Al's perception of the switchblade as a status symbol in his group.

Show in the interview the ways in which self-fulfilling prophecies might encourage different endings to each story as seen through the eyes of the three individuals. Indicate the significance of the ego states in the approaches of Al (adultlike), Johnny (childlike), and Mary (parentlike). How might the interview help each of the individuals grow in understanding of one another?

The incident of the switchblade may be somewhat more obvious in intent and effect than the next example. Bias becomes less distinct when concealed by self-righteousness.

Example 3-3
INTERVIEWER-CLIENT
MISMATCHING

Mrs. T. is visiting one of her clients at home. Her client's son has been skipping school and Mrs. T. has to discover the reasons for his truancy. Mrs. T. is not too pleased about visiting this client because of the dirty, run-down neighborhood. In fact, she wishes that someone else might have taken over this case. Of course, she is not prejudiced, but why has her supervisor assigned a white woman to deal with black clients? When she arrives at the drab and chipped door, she notices someone peeping from behind the torn curtain. Mrs. T. knocks on the door, carefully avoiding the rotting and splintered wood. "Not only dumb, deaf, too," she thinks. As she raises her fist to bang at the door a third time, she hears someone shuffling. The door is suddenly pulled open and a man in a dirty white shirt and baggy pants leans against the door. "Yeah, waddya wan'?" he slurs. Mrs. T. stares at the man and coldly asks, "Where is Mrs. M.? I want to speak to her." She thinks to herself, "This drunken bum. What's he doing here?" After a few more attempts to fulfill her assignment, Mrs. T. says, "We'll just have to get the truant officer here." Then she swiftly walks to her car and drives away.

Later, she describes the incident in her report to the supervisor, ending with the statement, "Uncooperative, unable to make contact. Man in the house. Welfare payments should be investigated."

Exercise 3-3
DISCLOSING SELF-RIGHTEOUSNESS

Write an interview between yourself and a client with whose appearance, opinions, race, or religion you feel annoyed or uncomfortable. Show by three statements that you feel that your feelings are justified, that you are "right" feeling this way. Then revise the content of the interview to show that you are striving for empathic understanding rather than to judge and disapprove self-righteously.

Discuss your interviews and compare them with other trainees. Do you still think, and do others agree, that you have succeeded in changing the tone of your interviews?

CHAPTER FOUR
FOUR DETERMINANTS INFLUENCING THE INTERVIEW

Developmental experiences. (Photo by Bruce Anspach, Editorial Photocolor Archives.)

When the client first enters the interviewer's office, he steps into a new world. The client enters the world of the interviewer, which is distinguished by all of the interviewer's values, feelings, and behavior, the effect of the interviewer's surroundings, and the attitudes and influences on the interviewer from the people around him.

The client brings with him the predominant culture of his parents, the way he feels about himself, the argument with his mother over his uneaten breakfast, the lack of money to buy a basketball, and the job refusal he received that morning. Along with the client arrive fears about the exam in psychology or whether he will "make out" with the new girl. The smile on the client's face reflects the memory of the exhilaration he felt when he managed to drive his car up the treacherous incline to the top of Pike's Peak on his recent trek across the country. The frown on the client's face was etched there from years of feeling that he is a failure and not knowing who or what he is or should become.

The interviewer also has memories of poverty or affluence, parental squabbles, and recognition or nonrecognition and introduces his style of humanity into the interview. Out of these determinants shared by the interviewer and the client comes the interview transaction.

BACKGROUND ROOTS

A search for information is not always necessary in each of the areas that, taken together, make up the background roots. Agencies differ about questions to be asked in the intake interview, and interviewers determine the amount of delving into the past, depending on whether they are oriented historically (past-oriented) or ahistorically (present-oriented).

Demographic factors

Demographic factors are those vital statistics that the U. S. Bureau of the Census gathers every 10 years. These statistical factors refer to the distribution of people in neighborhoods and certain characteristics of the inhabitants of each household. Such items as birthplace, race, sex, age, socioeconomic status, education, religion, and geographical origin furnish a cultural picture of the individual.

Physical factors

An individual's feeling about his body is closely related to cultural perceptions. Shakespeare recognized the influence of body structure on the perception of a person when he had Julius Caesar say, "Yond Cassius has a lean and hungry look; He thinks too much: such men are dangerous."

Attitudes toward well-rounded women differ among cultures. Hairstyles among different races as well as for men and women of the same race have fluctuated in length and spread. Physical factors affect the behavior of the people around an individual as well as the individual himself. These physical factors include body structure (large or small boned, obese or slim, tall or short), facial features, general state of health, medical history, and physical defects.

Developmental experiences

Both demographic factors and physical factors influence the opportunities offered to an individual. Where an individual lives as well as the extent of his movement beyond his immediate dwelling place becomes part of his life space. The narrowness of a ghettolike existence is contrasted by the freedom to explore.

Developmental experiences (what happens to and around a person as he grows up) provide the environmental opportunities. The degree of sharpness of the person's vision, hearing, and other senses opens or narrows his use of these opportunities. The developmental experiences that may expand or decrease the individual's potentials depend on his interpersonal relations with his family, neighborhood people, peers, and teachers. What happens to the individual at home, at church, at school, and in the military will affect his feelings about people, sex, and interpersonal relations in general.

Family history

The family into which the person is born (family of orientation) molds the future family style for the child become adult. The patterns of interpersonal relations practiced in the individual's family of orientation will be continued in a more or less modified fashion in the individual's family of procreation that he establishes after he sets up his own home. Factors such as parental relationships, parents or other significant adults with nervous or mental disorders, and the relationships of siblings (brothers and sisters) with each other are significant.

• • •

Listing the specifics of background roots increases awareness of their contribution to the creation of an individual's life-style. Exercise 4-1 provides a method of appraising the impact of background roots.

Exercise 4-1
DESCRIBING BACKGROUND ROOTS

Describe what *you* remember about yourself in relation to the areas that make up the background roots previously discussed. Then describe what *other people* have told you about yourself in relation to these areas.

Do you notice any difference in the recollections when you compare what you remember with what others have told you? If you do notice a difference, how do you explain it? Has the difference any relation to how you feel about yourself, your family, and others?

Make up a family system chart showing the people living in your home during your early childhood, during your adolescence, and at present. Indicate by the distance between the people you include in your chart how close or how far apart they seem to be to one another interpersonally. Place yourself on the chart, taking into account how important you feel in the family system as well as how close you seem to be to the various members. Place additional significant people not living within the home outside the square enclosing the family. Draw arrows or circles to portray the type of relationship between the people in your family (action/reaction, interaction, or transaction). Fig. 4-1 may assist you in drawing your own chart.

Does your family system chart give you any hints about how you feel about yourself and/or your family? In your description of your background roots, which events in your background seem to be most important to you and which of the people in your background seem to have had the most influence

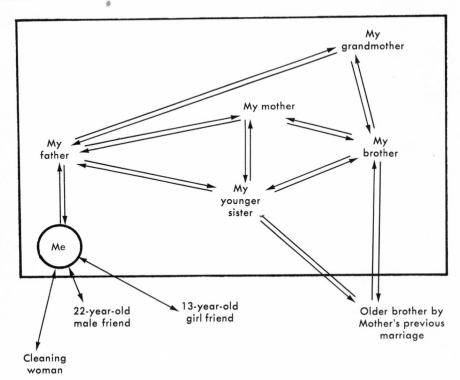

Fig. 4-1. Family system chart of 19-year-old woman who stutters and is depressed. She has attempted suicide within the last 5 years. Chart represents family situation 5 years ago.

on you? Do your background roots reveal anything about how some of your ideas regarding people have come about?

Examine Fig. 4-1 and describe what the diagram tells you about this individual.

Awareness of background roots serves another purpose. In the interview situation, characteristics such as the race of the client or the interviewer may provoke certain prejudices that may interfere with the flow of the interview.

Studies by Banks (1971) and others about the effectiveness of white interviewers with black clients indicate a slower establishment of rapport takes place than when the client and the interviewer are of the same race. Black clients reveal less information pertinent to the immediate problem when interviewed by white interviewers. This reluctance to be as free with a white interviewer as with a black may stem from a social conformity syndrome rather than from a racist attitude. This syndrome (pattern of behavior characteristics) is evidenced in the inhibition of hostility and the decrease in openness on the part of the black client. The black client may tend to act out what he considers to be the behavior expected by the white interviewer. This type of stereotyped interpersonal behavior leads to greater anxiety in the client and often a lowered performance on psychological tests (Heine, 1950; Baratz, 1967). However, studies (Yamamoto et al., 1967) have found that when the interviewer's level of prejudice is lower, the number of interview contacts with the black client is increased, and a more productive relationship is established

Butts (1972) describes the effect of black-white relationships from the viewpoint of the black psychoanalyst treating white patients. On the basis of his own analytical experiences with patients as well as those of other psychiatrists, Butts states that the black psychoanalyst with the white patient is afforded a unique opportunity for the following reasons. The analyst is able to examine the characteristics and changing expressions of racism of which the white patient may be unaware. Interracial analysis enables the white patient to work through certain feelings of inferiority/superiority that might not surface as easily when both individuals are of the same race.

Racial differences may induce patients to explore feelings against their own families that may resemble their racial stereotypes and hostility.

It is true that Butts' interpretation of the meaning of racism to his patients is based on psychoanalytical concepts of transference (the patient projects ideas, feelings, desires, and conflicts stemming from the past onto the analyst) and countertransference (the analyst does the same to the patient and uses the patient to gratify his needs). Nevertheless, whether the interviewer is psychoanalytically oriented or not, one should attend to four points in Butts' explanations.

1. Racial and other prejudices stem from one's past experiences in the family and in other social groups (background roots).
2. Interviewers may be effective in interracial interviews if they are capable of expanding their awareness of their racial attitudes and of controlling their reactions in the interview situation (psychological bases and situational features).
3. The interracial interview may provide the springboard for exploring feelings of prejudice and hostility toward people (situational features and psychological bases).
4. Racial differences or any differences such as age, socioeconomic status, and religious beliefs need not curtail expressiveness and sharing but may be used as sources for further understanding (background roots and behavioral effects).

The essential fact that should be gleaned from an examination of the effects of background roots is that an individual's attitudes toward any differences will affect the quality and quantity of attention involved in the interpersonal situation. Exercise 4-2 focuses on the discovery of attitudes toward differences.

Exercise 4-2
EXAMINING HANG-UPS AND HANG-INS
Hang-ups (dislikes and biases)

Write down all of the items of behavior or ideas or things that you find unpleasant and that annoy you. Concentrate on those characteristics of people or your relationships with people that "turn you off."

After you have completed your list, go back over it and rate all of the items on a scale from 1 to 5. Rate the items that bother you most 5, the least bothersome items 1, and the others in between. Gather together all of your 5 ratings and all of your 1 ratings. Look at these carefully. What do these lists of annoying items tell you about yourself?

Hang-ins (likes and preferences)

Write down all of the items of behavior or ideas or things that please you. After you have completed your list, go back over it and rate the items on a scale from 1 to 5. Rate the items that please you most 5, the least pleasing items 1, and the others in between.

Examine the results of both of your lists. What do they tell you about yourself? You might further your understanding if you compare your lists with others.

Exercise 4-3
A "HANG-UP" INTERVIEW

Select an item of behavior, an idea, or a thing that annoys you the most. Then choose someone to be the client who you feel may have the annoying characteristic or who will be most capable of role-playing the annoying client. Conduct a 5-minute interview with this individual, permitting the client to decide on the problem.

After the interview is completed, the interviewer, the client, and the observers should discuss the way in which the interview was conducted and the feelings that were revealed.

A second 5-minute interview should be conducted immediately after the discussion The group should then discuss whether there were any change in the interviewer's or client's approaches.

PSYCHOLOGICAL BASES

The background roots of the interviewer and the client stimulate the way in which these individuals perceive themselves and each other. These perceptions affect their levels of functioning and their expectations of one another. Out of these background roots, by means of maturation and learning, come certain psychological characteristics that are the bases for behavior.

As with the areas that make up back-ground roots, the areas involved in the psychological bases need not always be examined in their entirety. However, knowledge of the items and procedures for exploring them alerts the beginning interviewer to aspects that may be important in the development of the interview.

Self-concept

An individual's ideas and feelings about himself are a private matter. Each person erects his own cocoon of privacy, showing different parts of himself at different times and with different people. Some individuals reveal more of themselves and are often referred to as "open" people. Others may seem to be cold or withdrawn and are called distant or "closed" people. These closed people may hide distressing thoughts from themselves as well as from others. Penetrating their defensive walls may be difficult and sometimes dangerous since their defensive walls may provide a shield for more serious behavioral disturbances.

One way of looking at the degree of openness in the revealing of oneself is by means of the Johari window, a graphic model of interpersonal behavior. Joseph Luft (Luft, 1969) developed the concept of the Johari window with Harry Ingham (the term "Johari" being derived from the combination of their first names). The model lends itself to the understanding of any human interaction, including the interview process, since it takes into account the levels of awareness, what is happening within each individual (intrapersonal), and what is happening between the individuals (interpersonal). In addition, the "window" is flexible enough to permit examination of background roots and inner sources of behavior while viewing the ongoing processes and changes in the immediate situation (the interview or any encounter between people). Fig. 4-2 examines the individual self in a four-part division.

The person's *actual* self includes all of the attributes of the individual at any given moment. These attributes incorporate the psychological and physical characteristics developed from the person's background roots. Out of this total self the person's untapped and unfulfilled potentials (the real self) and the behavior, motivations, feelings, and potentials that are known to him (the

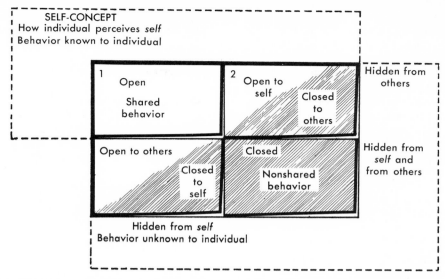

Fig. 4-2. The open and closed individual. (Adapted from Luft, 1969; by permission of National Press Books.)

self-concept) may be differentiated. Thus the self-concept encompasses the behavior, motivation, feelings, and potentials that are open to oneself and shared with others (division 1) and open to oneself but not shared with others (division 2). The self-concept excludes the behavior, motivation, feelings, and potentials that are open to others but not known to oneself (division 3) and closed to both oneself and others (division 4).

Furthermore, the degree of awareness of one's behavior, motivation, feelings, and potentials is unique for each person. The amount of revealed and concealed behavior may be dramatically demonstrated by changing the size of each of the four divisions.

In a new situation an individual is more likely to exhibit less open behavior by covering up what might be embarrassment or confusion with excessive smiling or other forms of politeness. This constricted behavior is more typical of individuals who are uncomfortable or feel inferior in social situations. In time the individual may open up more and thus demonstrate more shared behavior. However, the individual who has difficulty in relating or does not trust people is more likely to continue to close himself off by maintaining the same constricted activities with people or even becoming more withdrawn and uncommunicative (Fig. 4-3).

In general, it may be stated that the larger the area of shared behavior, the stronger the person's contact with reality. In addition, the person is more likely to use his abilities and make them available to others.

The question of openness, however, must be approached cautiously. Openness is not good in and of itself. Some people may be open (speak and act without restraint) as a defensive measure. For example:

Ted may believe that self-revelation will cover up the discomfort that he really feels with Jane. Sandy may reveal certain information about herself because she wishes to shock or verbally hurt Mark. Her hostile act may satisfy her for the moment but doesn't serve to further constructive growth in her relationship with Mark.

The open individual is committed to sharing ideas, feelings, and activities because of the mutual satisfactions involved in the interpersonal situation. The closed individual is committed to nonsharing to protect himself from real or imagined distress.

Robert Burns, the poet, was writing about behavior unknown to the individual but seen by others in his poem "To a Louse."

> Oh wad some power the giftie gie us
> To see bursels as others see us!

Mannerisms that may not be noticed by an individual are often caricatured by other

New situation

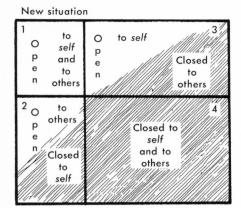

Changes in time

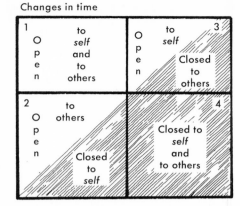

Fig. 4-3. Shifting self-concealment. (Adapted from Luft, 1969; by permission of National Press Books.)

people. The surprise and even shock of a teacher coming upon a student who is imitating the teacher's method of class presentation may range from mild distress to major disturbance. Characteristics of an individual become particularly noticeable when the individual is involved in social events, particularly when large groups of people are present. Some traits of which the individual may be unaware may include physical peculiarities in walk, body movements, gestures, speech mannerisms, or distinctive approaches in social situations that reveal the need to dominate or to be submissive.

The reasons for the lack of awareness of certain characteristics are complex. When these characteristics are just a part of the person's uniqueness, they present no impediment to the establishment of productive relationships. It is only when these actions interfere with satisfying one's goals that their discovery would be advisable.

When motives and feelings are open to oneself but not to others, this form of concealment is sometimes referred to as the result of ulterior motives. For example:

Bob may pretend that the only reason he wishes to tell Terry, his co-worker, about the boss's disparaging comments is to alert Terry to the hostility felt by Mr. M. Yet the ulterior motive may be that Bob hopes that such remarks will make Terry so uncomfortable in his job that he will resign. Then Bob will be able to move into Terry's job.

More and more lies become necessary to

continue this kind of concealment of motives. This deceit propels the concealing person into ever-increasing demands for more elaborate pretenses.

The most concealed area of behavior, motives, feelings, and potentials is that closed to oneself and to others. A great deal of energy is expended by the individual who forcefully forgets and conceals from others present thoughts or past experiences so painful that to face them would be destructive to the person himself and sometimes to others. This hidden area may contain behavior of which the individual is ashamed, acts he feels guilty about having committed or even those unfulfilled acts he has considered. There may be some less offensive events and beliefs that the individual has distorted to undue proportions. On the other hand, favorable characteristics may emerge. People in stress situations may surprise everyone by exhibiting unexpected leadership, peacemaker qualities, or great physical strength.

Expanding one's consciousness about oneself and about others often encourages better understanding. Exercise 4-4 is focused on discovery of self as seen by others and by oneself.

Exercise 4-4
HOW MUCH OF YOU SHOWS?

This exercise requires five participants. Two individuals talk to one another and the three others observe. One observer watches and listens to person A, another observer watches and listens to person B, and the third observer watches and listens to both person A and person B.

Person A and person B talk about any subject. This conversation should last 5 minutes. (If the participants prefer, they may assume certain roles and select a problem to talk about in an interview/client situation.)

After the conversation, a discussion is held for 10 minutes. First, person A and person B report any mannerisms and gestures they noticed in themselves and in each other. Then the three observers read from their notes and all five participants answer the following questions.

• How much of the observed behavior of person A and person B was reported by person A and person B as well as by each of the three observers? How might the differences in observation be explained?
• How much of the observed behavior of person A and person B revealed the feelings and thoughts of person A and person B? Did either person A or person B appear to be holding back? What cues of concealment did the observers notice?
• How much of the observed behavior of person A and person B reflected how person A felt about person B and how person B felt about person A?

In which of the four divisions of the "window" (Fig. 4-2) would person A and person B place themselves? In which division would the observers place them?

Person A and person B continue their conversation or role-playing for 5 minutes. During this conversation, person A thinks of something that he would like person B to do. This request should be an act that person A believes person B would be unlikely to perform because it is out of place in the immediate setting or because it does not seem to fit person B's way of behaving or some act that person A believes person B would consider ridiculous. Person A should try to manipulate the conversation in such a way as to induce person B to behave the way person A wishes him to.

After the conversation, the same procedure described in the first part of this example should be followed. Any differences in behavior between the first and second conversation should be noted.

Much of the way in which individuals view themselves is revealed in their aesthetic pursuits. The following poem is taken from *The Me Nobody Knows*, a book edited by Stephen M. Joseph (1969). Joseph, a teacher, gathered the writings of students in New York City. One poem boldly states a great deal about the self-concept of Frank Cleveland, a 17-year-old high school student, who used the pen name "Clorox."

What am I?

I have no manhood—what am I?
You made my woman head of the house—
   what am I?
You have oriented me so that I hate and distrust my brothers and sisters—what am I?

   You mispronounce my name and say I have
      no self-respect—what am I?
   You give me a dilapidated education system
      and expect me to compete with you—what
      am I?
   You say I have no dignity and then deprive
      me of my culture—what am I?
   You call me a boy, dirty lowdown slut—what
      am I?
   Now I'm a victim of the welfare system—
      what am I?
   You tell me to wait for change to come, but
      400 years have passed and change ain't
      come—what am I?

I am all of your sins
I am the skeleton in your closets
I am the unwanted sons and daughters in laws,
   and rejected babies
I may be your destruction, but above all I am,
   as you so crudely put it, your nigger.

Exercise 4-5
SELF-CONCEPT IN A POEM

After you have read the poem "What am I?" answer the following questions.
   1. What does this poem tell you about the background roots that influenced the way Frank feels about himself?
   2. How do you think Frank perceives himself?
   3. How do you think Frank thinks others perceive him?
   4. What prejudices for and against people, institutions, and ideas does Frank express?
   5. How much of Frank's feeling, thoughts, and behavior is he revealing, and how much is he concealing?
   6. How do you feel about Frank?

7. Do you think Frank needs help (counseling or therapy)? Support your answer by pointing to cues in the poem
   that led to your opinion.

These questions may be used for individual consideration but would be more
effective if used as springboards for small
group discussion.

Emotional status

Interwoven with self-concept are the
emotional status and the intellectual status
of the individual. Emotional status refers to
affective characteristics and intellectual
status refers to cognitive characteristics. The
emotional status of the client may be inferred
by the observed degree of emotional control
and of mood changes during the interview
as well as the descriptions offered by the
client of emotional behavior outside the
interview situation. The interviewer should
be keenly aware of the client's behavioral
responses to him. Some emotional characteristics revealed by the client's behavior
may be suspicion, reserved reactions, embarrassment, evasiveness, and ingratiation.

Intellectual status

Intellectual status or intelligence is one
of the characteristics of personality that
affects and is affected by the way the individual feels about himself. There is much
uncertainty as to just what intelligence is,
and therefore definitions of intelligence are
varied. Intelligence may be measured quanti-
tatively by means of standardized intelligence tests; often this type of measurement
implies a definition of intelligence as that
which the intelligence test measures. Intelligence may be measured qualitatively as
proposed by Piaget (1952), who defined intelligence developmentally according to four
main stages of increasing competence toward the achievement of logical thinking.
A third way to estimate intellectual status
is by means of the client's school and job
record and the vocabulary he uses. The
assumption underlying this third method of
assessing intelligence is that intellectual
capacity or cognitive ability is a tool used
for adapting and solving problems.

Determining the level of functioning of
both the client's cognitive and affective abilities is essential to helping the client. Frequently the intellectual capability of an indi-

vidual may not be demonstrated because of
overanxiety, lack of motivation, depression,
sweeping mood disturbances, or fear. At
other times an individual may evidence
intellectualization, a defensive maneuver
using intellectual activity in order to
conceal turbulent emotions or undesired
thoughts. When either rigid and/or blocked
intelligence or intellectualization is evidenced, the client's flexibility in confronting
new situations and in coping with his problems is reduced. Thus it is important to attain some measure of the status of the client's
intellectual functioning. The freeing of the
client's intelligence becomes an integral part
of the interview transaction.

One further caution concerning the assessment of intellectual status. The performance on most tests of intelligence is
dependent on cultural experiences and
especially on culturally biased verbal and
academic information. This cultural bias
makes these tests unfair to people from
different cultural backgrounds that stem
from racial, ethnic, economic, or geographical dissimilarities. People who do not have
the same opportunities as those individuals
for whom the intelligence test was constructed originally will be less likely to
achieve high scores on these tests.

Mental status

In the psychiatric interview the mental
status examination takes into account emotional and intellectual functioning and serves
primarily a diagnostic purpose. Mental
status assessment is usually administered
to the more severely disturbed individual
(the psychotic) whose degree of contact with
reality needs to be determined or when
organic brain damage is suspected because
of symptoms of brain injury or of senility
(aging process).

Mental status concentrates on the individual's current functioning. Although the information sought is recorded according to
subdivisions, the framework for gathering
the information should not be regimented by
the order of the categories of information.
Instead the sequence of topics and the flow
of conversation should be based on cues from
the client.

Examining for mental status requires a
great deal of skill that is acquired through
observation of a more experienced inter-

viewer and by means of practice under supervision. Skill is needed not only for the gathering of the information but also for the eventual organization and evaluation of the data in order to determine the degree of disturbance and the possible directions for treatment.

The information covered in the mental status examination includes preliminary data identifying the client and the source of referral; general description of the client; determination of the client's orientation, memory accuracy, and intellectual functioning; exploration of the client's affective level, reality orientation, and coping mechanisms; and the client's socialization level as evidenced in his interpersonal relationships.

SITUATIONAL FEATURES

The two determinants discussed up to this point are background roots and psychological bases. An individual's background roots build the foundation for *who* the individual thinks he is, his self-concept. Intermixed with the ideas, values, and attitudes that make up the self-concept are the levels of emotional control and intellectual responses. How much, when, and to whom an individual reveals himself depends on his life experiences (background roots) and certain features of the situation he experiences at the interview moment. Thus situational features are determinants in addition to background roots and psychological bases that influence the atmosphere and outcome of the interview transaction.

Many factors are established even before the interview begins. If the interviewer is in his own private office, the physical location of the office as well as what the client has heard about the interviewer are significant preinterview features. On the other hand, if the interviewer's office is located in an agency, the agency's functions and its physical location set the stage for the interview.

Agency functions
Functions defined

The purposes of the child guidance clinic, the family service agency, the community mental health clinic, the comprehensive health clinic, and the state or private hospitals vary in scope and intent. These dif-

ferences will affect the kinds of services provided as well as the particular focus taken for the problems.

Agency's priorities

The priorities set forth by the agency will also translate into such widely diverse aspects of the interviewing situation as to whether privacy and protection from interruption during interviews will be provided.

The interview that must be conducted in a doorless semienclosure has an atmosphere much different from one held in an enclosed room with a shut door. This lack of privacy may affect the client's responses and it may also reflect the agency's unfavorable attitude toward the interviewer.

Beginning professionals trained as human service workers, mental health technicians, mental health associates, or any of various titles used in such programs often find their status threatened in a subtle fashion. Acceptance by the agency and by the nurse, the social worker, the psychologist, the psychiatrist, and others becomes a complex process that is dependent on the positive self-attitude and degree of confidence and competence of the beginning professional. Acceptance and the development of the professional self were discussed in Part 1.

Physical location of agency

The physical location and appearance of the agency or interviewer's private office may have a significant impact on getting the client to come to the agency in the first place, and the various individuals in the agency or private office may encourage or discourage the client to continue the interviewing situation.

The storefront and the building in the inner city may be more acceptable locations for the client from the inner city. The middle-class individual who is impressed by distinguished addresses is likely to select and to be more open with the interviewer whose office is in an expensively designed and elaborately decorated building. For the status-seeking client a more sizable fee may be related to expectations for successful interviews.

Two ends of the continuum. . . . In one situation the simplicity and proximity of building and interviewer make for a favorable interview atmosphere. In another situation the affluent status of luxurious surroundings is assumed to enhance the helping capability of the interviewer. Physical location may have more or less of an effect on the outcome of the interview depending on the client's perceptions.

Interviewer's office

In addition to the amount of space and the degree of enclosure of the interviewer's office, other conditions contribute to the physical/psychological comfort or discomfort of the client.

Interruptions

The number of interruptions may have an effect on the conducting of the interview. The degree of disruption will depend on the frequency and length of these disturbances, how the client views the interruptions, and how the interviewer handles the interruptions.

Telephone calls and individuals that intrude into the office will disrupt the continuity of the interview even with the most experienced and competent interviewer. Perhaps the only value of such interruptions might be in providing the opportunity to observe how the client responds. A general rule that may prove effective is to not permit interruptions so that a free, uninterrupted flow of conversation may ensue. However, the interviewer must decide what emergency or other situation would permit an interruption.

Furniture arrangement

The privacy of the office may be increased by the physical arrangement of the office. Even in unsatisfactory circumstances the arrangement of the furniture as well as the type of furnishings will help to provide a semblance of privacy.

Whether the interviewer sits with a desk between him and the client or with two chairs facing each other is something the interviewer must determine on the basis of his own preference. A desk may prove to be a barrier for a feeling of intimacy; yet some interviewers are more comfortable behind a desk. The ease of relationship depends not only on the location of the desk but also on the ease of the interviewer.

Some clients may also get "uptight" with two chairs facing each other, particularly in the beginning stages of the interview. The positioning of the chairs and the kinds of chairs (straight-backed or upholstered) should be determined by concepts of comfort, genuineness, and respect as well as the feelings and needs of the client and the interviewer.

Other mechanics

There are other mechanics involved in the physical arrangements and furnishing of the office, such as the lighting and pictures. Is there too much or too little lighting? Is the sunlight or electric light shining in the client's or interviewer's eyes? Are the pictures on the walls provoking, calming, or stimulating? Are they appropriate for the particular clients most likely to visit the interviewer's office?

Other people to whom the client must relate in the agency

In addition to the interviewer, the client usually meets one or more other people to whom he must relate. The first individual the client meets, whether this person is the interviewer or other agency personnel, will become interlaced with the feelings the client has about the interview. For this reason some interviewers prefer to be the first person to whom the client speaks over the telephone or whom he meets in person. This first meeting is a matter of interviewer preference and is also related to whether the other people come on too strong or in some other way would disturb the client.

The brusque secretary or receptionist or other annoyed agency personnel may distress the client and consequently decrease the degree of openness the client feels in the interview situation. The attitudes of the client toward the individuals surrounding the interviewer in an agency or private office should be taken into consideration. Just as significant is how the people around the interviewer affect him.

Availability of other specialized services

In the fulfillment of the client's needs the availability of specialized services becomes an important issue. Ease of access to medication, specialized examinations, and other services make for a more effective and efficient helping situation. An adequate referral service is a must for the agency or interviewer in private practice where diverse services cannot be offered.

BEHAVIORAL EFFECTS

The contributing factors just described suggest two main concepts. One concept is that all behavior is caused; that is, there are stimuli (prompters) and reinforcements that perpetuate (condition) certain ways of responding and others that do not condition or extinguish other forms of behavior. Second, behavior is caused by many factors that are interrelated.

In other words, the client or the counselor learns his verbal and nonverbal behaviors. Furthermore, those behaviors that occur within the counseling transaction add another influence to the outcome of the session.

Example 4-1 describes the interrelationships among background roots, psychological bases, situational features, and behavioral effects.

Example 4-1
TED AND DR. L.
JUST DON'T FIT TOGETHER

Ted, 14 years old, is in the psychologist's office. One of the conditions of his release from the juvenile evaluation and referral facility was that he return home and receive psychological help. Ted had been picked up drug-wandering through the streets.

Ted is slouched in the chair opposite the interviewer. He just doesn't dig this guy who looks so much like his "damned father" with his straight suit and his straight haircut. The office fixtures turn him off. All furnishings are in an elaborate, expensive green-and-blue plan. They remind him of his own carefully arranged, affluent home. "Even the way this dude speaks is bookstuff. In fact, this guy's secretary sounds like Mother when she is trying to impress people."

Ted is weary of listening so he decides he will think about 2 weeks ago when Terry and he left the lunchroom at school and found Tricia for a fix. Ted looks at his hands and is determined that he will say as little as possible.

"Ridiculous," the interviewer, Dr. L., is thinking. "This character has everything going for him. I had to do it the hard way, working and studying at the same time. My parents couldn't afford to give me all the luxuries this kid has. They thought I should be earning money, not attending college. This pip-squeak just doesn't know what's good for him. My 5-year-old had better appreciate what I'm doing for him. Well, at least the fee is good. Let's see if I can penetrate this kid." Dr. L. briefly stares at Ted and then pulls his lips into a smile. He sits back in his chair, folds his arms across his chest, and begins speaking.

Ted looks up when he hears Dr. L.'s voice and mechanically mumbles his answers. Meanwhile, his right leg begins moving up and down rhythmically with a motion that seems to begin at the thigh. He slowly leans back in his seat and once more looks around the room, noting the diplomas and certificates on one side and the landscape paintings on the other wall. Ted thinks, "Even his walls are covered with lousy straight stuff." Then Ted's eyes move to Dr. L.'s face and he notices the thin bottom lip held by Dr. L.'s teeth. Ted slumps further into his chair.

Dr. L. untangles his arms and picks up a pencil for his desk. He moves the pencil back and forth between his fingers.

The example just described is exaggerated in order to condense several of the aspects of the four determinants. The ethical interviewer would refer Ted to someone else since the hostile responses of both the client and the interviewer appear to be insurmountable.

Summary of Part Two

In Chapter 3, three types of interview relationship have been distinguished—action/reaction, interaction, and transaction. The core difference between these three types of interviews is the degree to which the interview participants share the same interpersonal world.

In action/reaction interview, two separate individuals function independently with little or no concern for the impact of either their verbal or their nonverbal acts on one another. Interaction maintains this separation of individuals, but the interviewer's awareness of his effect on the client becomes an essential aspect of the interview procedure. The interviewer uses the return messages from the client (feedback) to alter the course taken toward interview goals. With an increased awareness of the interaction process, the interviewer is alerted not only to his effect on the client but also to the client's effect on him. Thus in transaction the interviewer and client maintain their identities and yet are intimately involved in a social system consisting of two people (dyad) or sometimes more who are sharing an interpersonal world. Both the interviewer and the client change and grow in understanding in the transaction interview.

Since the transaction interview resembles the process of transactional analysis, a comparison of the similarities and dissimilarities of these two aspects of the helping process was made. The transaction interview and transactional analysis are similar inasmuch as both stress the social system established in any interpersonal relationship as well as the active inclusion of all participants in the social system. Dissimilarity stems from certain transactional analysis concepts, for example, the three ego states of parent, adult, and child, which are not utilized by all interviewers in their transaction interviews.

As discussed in Chapter 4, background roots, psychological basis, and situational features all in one way or another affect the behavior of both the interviewer and the client during the interview. A dynamic transaction results from all of these determinants and the impact of each varies at different periods of the interview. At the beginning, perhaps, the background roots and situational features may be most influential. Gradually psychological bases become predominant in interaction and behavior. All determinants are more or less present at all times.

During the interview, both the interviewer and the client change. Whether the change is toward greater fulfillment of both depends in part on the background roots, psychological bases, and situational features, but particularly on the skill of the interviewer.

PART THREE

ALL INTERVIEWS ARE DIVIDED INTO THREE PARTS

HELPING THE CLIENT FIND HIS WAY

CHAPTER FIVE
VARIETIES IN INTERVIEW GOALS

Social reinforcement. (Photo by Bruce Anspach, Editorial Photocolor Archives; from Poland R. G.: Human experience: a psychology of growth, St. Louis, 1974, The C. V. Mosby Co.)

The degree to which an interviewer focuses on background roots, psychological bases, situational bases, or behavioral effects evolves from the primary goals of the interview. Three general varieties of interviews will be discussed in this chapter—information oriented, behavior modification, and experiential.

INFORMATION-ORIENTED INTERVIEW

The information-oriented interview has the least depth in regard to interpersonal searching. Two variations of this orientation are discussed.

Information-seeking interviews

The information-seeking interview is directed to gathering data about people, places, or products. The interviewer seeks from the client facts or opinions about some predetermined items.

Survey interview

The telephone rings and one more opinion about television is sought. "What program are you watching now?" "When did you begin watching?" "How many television sets are there in your home?" The survey takes many forms—preferences in TV programming, in food packaging, in magazines bought and read, etc. Usually the queries serve to inform the television sponsor or the marketing agent how to increase his product's appeal to the consumer.

Public opinion poll. A survey may center on public opinion. For instance, the Gallup Poll asks carefully planned samples of the population selected questions about their presidential candidates, their political opinions, as well as other questions. In the January 30, 1973, issue of *The Washington Post*, George Gallup described the results of a nationwide survey he conducted after the announcement of the Vietnam peace settlement by President Nixon. Gallup announced the percentage of people whom he polled who were satisfied, dissatisfied, or undecided about the nature of the peace agreement. Opinions about "peace with honor" and some pessimism about the future situation in Vietnam were examined by means of questions that required a reply of "yes," "no," or "no opinion." The questions in a Gallup Poll are direct, closed questions very much like multiple-choice questions in a classroom test.

Personnel interview

Another sort of information is sought when an employer is reviewing the qualifications of an applicant for a job. The suitability of the candidate is determined by a discussion of his educational background and experience. Often the applicant is asked about his interests and what makes the applicant believe he would like and can successfully perform a particular type of work. Certain hypothetical incidents may be posed and the applicant asked to discuss how he would handle them. On the basis of such observation and questioning an applicant is selected.

Personnel interviewers in larger organizations frequently go beyond the job placement functions of preliminary recruitment, screening of applicants, and recommendations of the more qualified candidates to some form of counseling. Thus these interviewers expand their functions to a helping relationship.

Journalistic interview

The reporter who makes inquiries for a newspaper story must have at his fingertips a large selection of questions to probe for information that he can organize into a story. The reporter seeks facts to cover the five W's: *Who? What? When? Where? Why?* In writing his copy the reporter answers all these questions in his lead paragraph. In addition, the interviewer may add a *how* question to his inquiry. News and feature reporters must be skilled in interviewing people to collect interesting and accurate information for their articles. The successful interviewer picks a knowledgeable interviewee, asks the right questions, and knows where to begin.

• • •

In order to obtain information for the information-seeking interview, the interviewer must establish objectives and questions leading to the achievement of these objectives. These questions may be presented orally or the client may be requested to complete the answers on a questionnaire.

Testing

The tests that often plague students are forms of an information-seeking interview. The information (answers to questions) sought by the interviewer (the instructor) from the client (the student) depends on whether regurgitation (spilling back what the instructor has said) or self-understanding and interpretation (student oriented) are desired.

Information-sharing interview

The information-sharing interview goes a step further than information seeking since the data gathered is used for more than achieving the interviewer's own goals. Information sharing serves to help the client make a decision, for instance, about educational or vocational plans.

The ultimate purpose of this information sharing is similar to the diagnostic interview in which the function of the sharing is to provide both the interviewer and the client with information about the causative factors and the procedures needed for therapy. For the physician, the clinical examination of the patient, the x-ray films, the electroencephalogram, and the stethoscope help in determining possible factors involved in the patient's illness as well as the seriousness of the illness. *How much* and *how* this information is shared with the patient becomes a matter of the physician's particular approach and the physician's evaluation of the patient's needs. The mental status examination discussed in Chapter 4 is an additional illustration of information seeking that may lead to sharing. Example 5-1 represents a psychiatrist's approach to a mental status examination of a recently hospitalized woman who is 62 years old.

Example 5-1
INFORMATION SHARING IN THE MENTAL STATUS EXAMINATION

Ms. P. enters the psychiatrist's office slowly. She walks with her head down but moves her head up and to one side as she quickly stares over her shoulder. She slips down into the seat that faces the psychiatrist, keeping her head bent with her chin held tightly downward. Her hair hangs loosely over her forehead and the hem of her flowered dress droops down on one side.

The psychiatrist, Dr. L., starts to say "hello" but gets no further than "He. . . ."

MS. P.: *(Rapidly raising her head and looking backward over her shoulder.)* Hear it? It's again. Again, again! *(She looks with upturned eyes at Dr. L. and then quickly pulls*

her head down.) (NOTE: Auditory hallucinations and poor reality orientation.)

DR. L.: No, I did not hear that this time. Tell me about it.

MS. P.: Hear it? Tells what to do. Since I came here 10 years ago. (NOTE: Present memory inaccuracy.)

DR. L.: Ten years ago. How old were you then, Ms. P.?

MS. P.: Forty-two—no, maybe 44. Don't remember. Ten years ago came here to this hotel.

DR. L.: Hotel? Ms. P., this is a hospital. Do you remember that your daughter and your husband brought you here?

MS. P.: *(After a few minutes.)* So tired today. That new medicine gave me on Tuesday.

DR. L.: Medicine on Tuesday?

MS. P.: *(Pulling her eyebrows together and wrinkling her forehead.)* Must have been in the morning yesterday, Wednesday, or Thursday. Hard to remember. *(Raises her head suddenly and looks over her shoulder.)* Again! (NOTE: Difficulty with present and past memory accuracy.)

DR. L.: Tell me about it. Where does the sound come from? (NOTE: Testing for reality orientation.)

MS. P.: Have arranged for an investigation by the F.B.I. and a supersonic witness. Must reconstruct the ions to induce closure. Have to go now. More to do. Remember too much. (NOTE: Vocabulary level high and intellectual functioning poor.)

Ms. P. appears to have some difficulty raising herself from the chair. After she manages the initial spurt of movement, she pulls herself up and slowly walks from the room. Dr. L. notices a slight shuffling limp in her left leg. Dr. L. follows Ms. P. to the doorway. He looks out and beckons to the woman seated on the wooden bench outside his office.

DR. L.: Please come in, Ms. R. Now that I have spoken to your mother, we can talk about the questions you asked the other day. As you said when we talked, your mother is an intelligent woman but is not functioning in accordance with her intellectual capacity.

MS. R.: Doctor, do you think she's ever going to be as she was—an alert, active woman?

DR. L.: We need some more tests to determine her prognosis. There are some evidences of brain pathology, but I'm not certain enough to state the nature of the pathology. A thorough physical examination, including an encephalogram, is indicated.

Dr. L. discusses the patient's poor reality contact and hallucinations as well as his other findings. In this information-sharing interview the data gathered about a person is shared with someone other than the patient.

Exercise 5-1
DISCOVERING THE MESSAGE IN THE INFORMATION-SHARING INTERVIEW

Dr. L.'s interview with Ms. P. not only opens up many areas for further exploration but also offers many cues. Examine the interview carefully and answer the following questions:

1. What additional information does the doctor require to obtain a more complete picture of Ms. P.'s problem?
2. How might Dr. L. have obtained more information and a better understanding of Ms. P. during the interview?
3. On the basis of the information presented during the interview, what inferences would you make about the causative factors of Ms. P.'s behavior and the diagnosis?
4. If you were the interviewer, what would you have changed in the interview with Ms. P. or with her daughter, Ms. R.?

BEHAVIOR MODIFICATION INTERVIEW

The behavioral counseling or behavior therapy interview is both symptom removal and action oriented. The techniques to be followed depend on which theories and principles of behavior influence an interviewer upholds. In general the formulas of behavior modification are founded on the groundwork of experimentally discovered principles of learning. Basic to the practice of behavior therapy is the hypothesis that both appropriate ("normal") and inappropriate ("abnormal") behavior are learned and therefore may be unlearned. Inappropriate behavior is weakened and eliminated and appropriate behavior is instituted and strengthened by means of schedules of reinforcement (reward

or punishment). This process is called operant conditioning or contingency management. (See the ABC of recording in Chapter 2.)

Operant conditioning is defined as emitted (voluntary) behavior that is repeated (conditioned) when the reinforcements (rewards and punishments) are consequences of the behavior. Thus the behavior is operant because it *operates* on the environment. When an individual comes to a locked door, he puts a certain key in the lock, turns the key in a certain direction, turns the doorknob, pushes the door open, and walks into the house. The individual's behavior has been shaped to perform several operants (actions) that have now become habits or persistent behavior: to discriminate stimuli—the appropriate key for the lock, the appropriate way to turn the key; to discriminate responses—move his arms, fingers, and legs in the appropriate way; and to maintain neuromuscular control—use certain muscles in opening and shutting the door. All these acts of behavior have been chained together by the reward (reinforcement) of entering the house. Entering the house is the contingency, the condition that happens at the right time and the right place, to reinforce the behavior of using the key to unlock the door.

The behavior modification interview is based on the concept of consequences or reinforcements of behavior. Behavior may be conditioned so that it continues or is extinguished by the nature of the reinforcement program.

Behavioral analysis

The first step in deciding the procedures to be followed for behavior modification is to define the target behavior or behaviors to be eliminated. As the interview progresses, the terminal behavior to be strengthened evolves. The behavior that is the target to be changed is reviewed with the client to determine the cues (stimulating or antecedent events) and consequences (reinforcements) that perpetuate the behavior.

The behavioral analysis of the obese individual who gorges himself on rich food leads to data about the frequency, quantity, and types of foods that are eaten. The interviewer and the client plan a "behavioral contract" (Krumboltz and Thoreson, 1969) that spells out a program of reduction of food intake and the specific rewards doled out. The rewards may be the weight loss indicated by a scale, the interviewer's social approval as well as the approval of the client's family and friends, a smaller size in clothing, or any other forms of reinforcements applicable to this particular client. Further reinforcement for change may come from snapshots of before, during, and after the weight loss. Thus signs of progress may be shown in snapshots revealing the flattening of the bulge. Negative reinforcement may be sustained by the client between interview visits by looking at the "before" snapshots (self-monitoring contingency) when sitting down to a luxurious spread (self-regulation or self-management). Furthermore, the stimulation to eat fattening foods may be controlled by limiting the cues around the client—removing any such food from the client's home and substituting less fattening food and drinks for candy, cake, and bread.

Reinforcement menu

After the target behavior has been determined (less food intake) and the terminal behavior decided on (maintaining a certain lower weight), the other parts of the reinforcement menu are considered.

Thought stopping

Wolpe (1969) writes about obsessive thoughts that may be diminished by a "thought-stopping program." The client is told to close his eyes and to talk out loud about the continuing, annoying thoughts that may include food, sex, writing a book— anything. At certain intervals (intermittent reinforcement) the interviewer sharply calls out, "Stop!" (negative reinforcement). The client's eyes snap open and the interviewer calls his attention to the change in thought (annoying thoughts suddenly stop) that has occurred. This procedure is repeated several times.

The interviewer suggests that the client close his eyes again and this time talk about something that is pleasant (substitute behavior) to him. If the client returns to obsessive thinking, the interviewer once again utters his loud command, "Stop!" In addition to what happens during the interview, the client is given the directive to practice stopping his disturbing thoughts by quietly reminding himself to stop. Extinction, or the gradual weakening of a response, is thus accomplished by a schedule of both negative ("stop") and positive (interviewer's approval) reinforcement.

Interpersonal characteristics

The interviewer himself is part of the reinforcement environment. Truax (1966) speaks of the differential reinforcement of the interviewer's high levels of accurate empathy, nonpossessive warmth, and genuineness. These interpersonal characteristics set up a reinforcing climate that supports and strengthens the client's self-exploration efforts, which lead to consequent changes in his behavior.

Verbal and nonverbal conditioning

Besides the three interpersonal characteristics, there are at least two other forms of reinforcements that are always part of the interview setting. These are verbal and nonverbal conditioning (Woody, 1971). Verbal conditioning refers to the words the interviewer uses that influence the client's responses. The interviewer smiles or nods his head or performs some other nonverbal act when the client performs some specified act or statement. Conversely, the interviewer ignores (avoids responding) when the emitted act or statement has not been specified. By means of verbal and nonverbal reinforcement, appropriate behavior is strengthened and inappropriate behavior is weakened. It is therefore essential that the interviewer be alert to his own responses as well as to the client's for the establishment of an efficient and effective reinforcement menu.

From this discussion the controlling power of verbal and nonverbal reinforcements becomes apparent. Not only the client's behavior but also the interviewer's behavior is modified by verbal and nonverbal conditioning. In fact, this is an everyday occurrence. The mother calls out to her child, "Stop that and come here right away!" Whether or not the child does come right away depends on the consistency of consequences for coming or for not coming. Sometimes it becomes more reinforcing (more fun and rewarding) not to come because then the mother will come into the room and thus give the attention that the child desires. Attention is the reinforcement.

Desensitization

Another circumstance in which attention modifies behavior is when the interviewer helps the client get down to his gut anxieties. By attending to the anxieties, by talking them out, the client becomes increasingly desensitized to the anxieties. When the strength of the anxiety-provoking stimuli is not sufficiently reduced by means of exploring the anxiety, then Wolpe (1969) describes the behavior therapy procedure that is based on the construction of a hierarchy of anxiety stimuli. This hierarchy is planned so that the least anxiety-producing stimuli are tackled first and by gradual steps the more anxiety-producing stimuli are attempted. Throughout this process of extinguishing anxiety and fear responses, the reinforcing relationship with the interviewer is crucial. Example 5-2 shows how desensitization may be used with a client whose anxiety is produced by speaking before a group.

Example 5-2
DESENSITIZING SPEAKING
ANXIETY

This was Sally's third visit to Mr. R. She thought about the last two sessions as she walked along the corridor to his office. "Guess I really dig the guy. Didn't when he pushed himself into my life at first. At any rate, thought he did. When, where, how—dig a little, dig some more. Now, can do two of the things we planned. Talk to two people at the same time. Was that a drag! In classroom think about what will say but don't feel have to answer. That was easier, yet at one time would have been so tied up, felt had to answer, wouldn't have worked. Sit and think—know my answer is mine to give if I wanted to. Well, here goes number three."

Mr. R. was in the anteroom as Sally entered the office. He looked toward her, smiled, then gestured toward his office door, and followed Sally as she went in. They both sat down in the facing chairs.

MR. R.: Hi, Sally. Hey, what's this? Looks to me like you're uptight today. Let's slump a little, Sally.

SALLY: Cut it, Mr. R. Just thinking about number three.

MR. R.: Number three?

SALLY: Yeah. Don't think I'm ready for it. Speechicating to you about anything.

MR. R.: Speechicating—excellent word.

SALLY: Planned it. Have been thinking about what we've done—the tight knot in my stomach was not there with the first and the second, but now.

MR. R.: *(Smiling.)* Planned it.

SALLY: Yes, I've been thinking about the meaning of some of the games people play, you know, Berne's (1964) book. I see how it applies to me—the child afraid to speak. The games I play, the "If-it-weren't-for-you" game; my parents did it—my boyfriend does it—they are the ones that stop me from doing things. "If it weren't" . . . a nice game to hide in. Whaddya know . . . here I've been rambling, speechicating, and am unknotted.

Mr. R. smiled and nodded his head and Sally continued her discussion of Berne's ideas.

Exercise 5-2
ESTABLISHING ANXIETY
HIERARCHY

Use the example of Sally's difficulty as the starting point for the following activities.

Set up a small group of four people. Allow 10 minutes to plan the steps in the anxiety hierarchy that will lead to the fulfillment of Sally's terminal behavior of speaking before a group.

Two members of the group role-play a 5-minute taped interview showing how Sally's attempts to achieve step four may be reinforced. The other two members of the group take notes on all of the verbal and nonverbal reinforcements used by both interviewer and client.

The tape is played back and all four members of the group take notes on the verbal cues and reinforcements. The effectiveness of the reinforcements is discussed and suggestions for improvement are made.

The same two group members resume their roles and conduct a 2-minute interview that is taped. After playing back the tape and noting the cues and reinforcements, the four members discuss the improvement in the effectiveness of the reinforcements.

Items on reinforcement menu

The following lists give some examples of appropriate and inappropriate behaviors that are modifiable by reinforcement contingencies and examples of negative and positive reinforcements.

Appropriate behaviors to be strengthened include the following:

1. *Communication.* Enunciation of comprehensible sounds or words
2. *Use of materials.* Appropriate use of blocks for building, not for striking another child
3. *Following directions.* Learning to comply with requests and to follow certain procedures in becoming desensitized to certain fears
4. *Social development.* Reinforced to make voluntary movement toward social encounter with another individual or to talk to another individual

Inappropriate behaviors in the same areas that should be weakened include the following:

1. *Communication.* Reinforcement schedule with planned steps toward extinguishing stuttering
2. *Use of materials.* Prohibiting use of the record player (object removed) for 1 day because player was roughly handled
3. *Following directions.* Removal of some privileges due to smoking in bed
4. *Social development.* Tantrums ignored or negatively reinforced by isolation from group, for instance, in a time-out box

Reinforcement is most effective if it is tuned in to the interests of the client. Negative and positive reinforcements are therefore relative terms depending on the client for their impact. In general, one or more of the following reinforcements are used in the conditioning process.

Positive reinforcement	Negative reinforcement
1. Use a particular food preference as reinforcement.	1. Remove food.
2. Give tokens in certain amounts for specified behavior. Tokens may be of wood, metal, paper, or actual money.	2. Charge fines of certain number of tokens for nonspecified behavior.
	3. Express social disapproval in nonverbal form—frown or movement of

3. Provide social approval in nonverbal form—facing the client, smiling, or a pat on the shoulder. Verbal compliments such as "You did it well" are also effective social reinforcements.

head from side to side ("no") or verbal onslaught such as "You're not really trying." Ignoring or not looking at client is another form of negative social reinforcement.

Assertive training

Modifying behavior is a many-sided phenomenon that takes into account the principles of learning, the client, the interviewer, and the environment in which the client lives. The client must learn to cope with more than the environment within the interview office. In order to accomplish this mastery of the environmental effects on his behavior, the client must learn self-assertion. The interviewer advances this behavior change by reinforcing positive features of the client's self-image (self-concept) and by establishing a more flexible and more satisfying self-reinforcement system.

The task of behavior change can be simplified if the persons and events in the client's natural environment are programmed to respond so that these factors reinforce the client's appropriate behavior and positive self-concept. The husband who approves of his wife's drinking by subtle reinforcements such as "I know you're trying, but if you don't stop right away, I'll understand" is really part of her problem of alcoholism.

Another way out of this dilemma is to condition the client so that he is better able to cope with the direct and indirect reinforcements in his environment that are perpetuating his unsatisfying (to him) and inappropriate (to others) behavior. Four concepts must be clarified for the client to learn to cope with his environment more adequately.

Differences between assertive, nonassertive, and aggressive behavior

Each person has the right to express himself but also has the responsibility to facilitate the personal growth of others. Differentiation between expression (assertion) and aggression (unbridled release) is essential to produce a favorable atmosphere for fulfillment of self and of others. Dealing with these

Table 2. Nonassertive, assertive, and aggressive behaviors*

	NONASSERTIVE	ASSERTIVE	AGGRESSIVE
Self-expression	Self-denial; others feel sympathy or sometimes contempt toward this person	Enhances feelings of self-worth as well as self-worth of others	Enhances self but denies worth of others; others feel put down and even humiliated
Exhibiting feelings and/or acts	Does not feel free to say or do what is reasonable or right; conceals feelings because easily hurt and wants to avoid criticism	Expresses what he feels and thinks and encourages others to do so; usually has friendly, affectionate, nonanxious feelings and is sufficiently self-assured to accept or reject criticism	Expresses what he feels and thinks but denies opportunity for such expression by others; usually feels resentment and anger and considers criticism unjustified
Selecting path of action and/or decision making	Easily maneuvered into undesired situation because of compliant surface behavior; fears rejection or ridicule of own decisions	Chooses in accordance with own needs without infringing on rights of others to choose	Usually chooses only in accordance with own needs, disregarding effect on others
Reaching goals	Often stops short of goal or doesn't try too hard to achieve it; feels inadequate to task and lacks self-confidence; others may push ahead of him, achieving his lost goal	Usually achieves goal and encourages others to achieve their goals	Achieves goal by downgrading others; thus others usually do not achieve goals
Feelings	Feels hurt, crushed, anxious, self-conscious because does not assert himself; others may feel guilty or angry at incompetent behavior	Assertive responses make him feel less anxious and more competent; others feel comfortable and at ease in expressing themselves to him	Depreciates others so that they feel hurt, defensive, and incompetent
Responses to unreasonable request	Probably concedes to request; others may feel guilty and annoyed with his inability to stand his ground	Politely refuses and explains reasons; others may not be happy about refusal but usually feel there is some justification for his refusal	Refuses angrily, which makes others either defensive or angry
Attitude toward origin of difficulties	More often is intrapunitive (blames himself for whatever goes wrong) and punishes himself (negative self-reinforcement)	More often is impunitive (free from punishing responses); more likely to rationally evaluate events	More often is extrapunitive (blames others for whatever goes wrong) and tries to punish others (negative reinforcement for others)

*Adapted from Alberti and Emmons (1970).

differences and knowing when and how to stand up for his rights is often difficult for the client.

The Western culture encourages the individual to move toward nonassertive or toward aggressive behavior. The quiet, obedient child is often the most praised and accepted at home and in school. When the child is impelled to exercise his feelings and his rights, he often learns to be aggressive rather than to assert himself. Unfortunately the consequence of aggression increases anxiety. Assertive behavior, however, acts as an anxiety decreaser.

When the interviewer helps the client to become assertive, the client becomes the master rather than the victim of his disturbing symptoms. Table 2 indicates the dissimilarities between nonassertive, assertive, and aggressive behavior.

Establishing the baseline

Baseline data refer to the frequency of the behavior to be changed. Such data provide information of where to begin in the conditioning process and how serious the behavior is for the client and for others around him. In order to obtain baseline data, several leads must be followed: (1) the extensiveness of the behavior under study, (2) the specific situations under which the behavior occurs, and (3) the recognition of the inappropriateness of the behavior.

Extensiveness of behavior. The extensiveness of the client's nonassertive and aggressive responses is one of the leads to pursue. The accumulation of evidence of inappropriate behavior frequently supports the picture of an oversocialized client who has been conditioned to the underlying position of nonassertive behavior or to the upperhand status of aggressive behavior. If the life history of the client were examined, the selected reinforcements discovered would reveal that as the child grew into adulthood the reinforcements overemphasized social obligations. Thus the rights of others were embellished with greater vitality than were the client's rights.

Dan illustrates nonassertive behavior issuing from this distortion of obligations. He tells the interviewer about what happened just last night when he was waiting in the line to buy a ticket to "Slaughterhouse Five." Some fellow ran alongside the line, stopped suddenly next to Dan, and elbowed himself into line in front of Dan. Dan's impulse was to "push the guy out"—but he didn't. Instead, Dan comments about his feelings of "uneasiness" and "embarrassment" because, as usual, he proved himself "incompetent in standing up" for his rights.

"Besides, Mary was annoyed with me," says Dan as he raises his shoulders and slowly drops them. "Mary is not a silent woman. She bristles when she doesn't like something—no one easily puts something over on Mary." Dan continues to explain how Mary's yelling at the guy just made him feel smaller and more defeated, particularly when the guy looked back at them and smirked.

From Dan's description, one may assume that Dan is generally nonassertive while Mary is usually aggressive in response to distasteful social encounters. Some individuals do not assert themselves or become aggressive in most situations; others are nonassertive or aggressive with certain people and/or with certain groups of people. Alberti and Emmons (1970) discuss the differentiations among general nonassertiveness and aggression and situational nonassertiveness and aggression.

The extent to which an individual bends to the domination, exploitation, and offensiveness of one certain person, several people, or most people hinges on the degree of the individual's positive or negative self-esteem. The lower the level of self-esteem, the more widespread is the individual's restricted expression of his feelings and the assertion of his rights. The choice of behavior, nonassertiveness or aggression rather than assertiveness, depends on what works, that is, what kind of behavior has been reinforced in the past and is still being reinforced in the present.

The interviewer helps the client sort through the maze of circumstances to discover the instances when the client is apt to inhibit actions, words, or exhibition of feelings. In order to determine the extent of restraint the client is encouraged to talk about the scope and strength of these social chains with which he has locked himself. The interviewer observes the client's verbal and nonverbal signs of anxiety as he erects a pattern of "people walk all over me," "everyone gets ahead of me," "can't get anything I want," "constant battle to get anywhere," "claw and scratch my way through life," or "there's a sucker born every minute; why not take what you can get?" Discovering the specific or generalized particulars of the inter-personal anxiety that results in either de-creased or belligerent action assists the client and interviewer in planning a behavior mod-ification program based on reinforcement of assertive behavior.

Specificity. Several factors in addition to individual life experiences induce an indi-vidual to respond with more assured asser-tion in one situation than in another. The different reinforcements prevailing in West-ern culture have regulated masculine and feminine behavior. This is not only evidenced in distinctive mannerisms but also in the exercise of assertion.

In measuring the meaning of the client's behavior, cultural impositions must be taken into account. This would hold true not only for the reinforcements for behavior accord-ing to sex but also in regard to subcultural distinctions originating from racial and eco-nomic expectations. Thus what may be characterized as assertive behavior in one social situation or social group would come across as aggressive or even nonassertive behavior in another situation.

The translation of the script that contains the client's behavior depends on the social situation from which the client comes and the degree of rewarding consequences open to the client because of his behavior. Armed with data about when, where, and how, the client takes on the tasks of social living and self-fulfillment.

Recognition. Before any behavior change can be effected, the client must realize the inappropriateness of his behavior. He may feel uncomfortable and want events to turn out differently, but he may not be fully aware that it is what he has been and is doing that is important. This returns to the concept of self-fulfilling prophecy. It is almost inevitable that the client is going to be a loser. He not only looks and acts the part of the loser, but he manages to get him-self into situations in which a loser is the only part that is left for him in the drama. He must begin to notice the bad smell of the place he is in and sometimes leaving the

scene is the most assertive behavior he could portray.

Others may help or hinder the client in continuing a certain character part. Another factor in the role-continuation is the client's reaction and his ability and/or desire to institute a new action. Sarah exemplifies this difficulty to note one's own self-reinforce-ment of a part in a self-written script.

Example 5-3
SARAH WRITES HER SCRIPT

Sarah trudges through the waist-deep snow and feels her hand getting stiffer. Soon she doesn't even feel her left hand and the initial pain is gone. She walks into the class-room, drops her books on the floor, and sits down. Then she begins to feel an uncom-fortable tingling sensation in the tips of her fingers. She thinks to herself, "Defrosting, that's what's happening. Hell, it hurts." She gets up from her seat and with a brief look at the professor up front, she runs out of the room.

"Stupid," she says to herself, "Stupid, why didn't you say something? He's going to think you're some kind of a nut. What a jerked-up thing to do." She walks as rapidly as she can to the nurse's office, looking straight ahead as she walks. Some of the people walking by look at her and smile or say hello, but since Sarah does not respond, they walk away.

Sarah enters the nurse's office and explains to the nurse, Ms. D., "Guess my hand is frostbitten. Is there anything that can be done about it?" Ms. D. puts some warm packs on Sarah's hand and, although the severe pain of returning circulation as the hand thaws is tremendous, Sarah speaks to the nurse about the depth of the snow and how unusual the storm is. When Sarah re-covers the use of her hand, she is careful to thank the nurse and to commend her on her assistance.

As Sarah leaves the nurse's office, she realizes that she feels unusually tired; as she walks along the corridor to the classroom, she notes that she sees some dancing black spots. "Oh, no," she says, "that's all I need now, a headache." A slow throbbing begins in her right eye and Sarah knows that she is going to have a migraine. "Why didn't I say something to Ms. D.? Could have gotten some pills. Began to feel something then, that horrible gnawing feeling in my stomach."

"Nut, nut, stupid nut. Didn't say anything to him [the professor] or her [the nurse] . . . all show. Don't want anyone to know. Damn, it's getting worse. Or is it I'm making it worse? Wow! Do I feel sick, nausea, so tired, and *have* to go back into the room. Everyone is sure to look, yep, stare at me. They'll think, 'There she goes again, the snob in bo-peep clothing.' Bo-peep? What do I mean by that? Why do I always do things the wrong way? Speak up . . . say up . . . throw up. Can't give forth my feelings. No wonder Joe called me a 'cold fish' the other day. Nut, nut—what a nut. Throbbing—oooh! The throbbing, splitting my head. False pride, he said. Oooh!"

Exercise 5-3
REWRITING SARAH'S SCRIPT

Arrange small groups of four participants. Each individual in the group should select one of the following roles: Sarah, the nurse, the professor, or the counselor. For 2 minutes each the nurse and the professor tape what they think and feel about Sarah.

Role-play and tape a 10-minute interview between Sarah and the counselor.

Sarah has come to the counselor about her inability to show her feelings. This is the only problem she relates when she first enters the office. Since the counselor knows Sarah is coming to see him, he pulls out her medical record from the college health services. The counselor notes she has a history of headaches, which she dates from when she reached puberty at the age of 14. Conduct a 10-minute interview during which the counselor shows how he begins his interview, starts exploring Sarah's concerns, and discovers how and under what circumstances Sarah handles social situations assertively, nonassertively, and aggressively. The counselor should integrate what the nurse and the professor say about Sarah in his interview.

Listen to the tape and then for 10 minutes discuss Sarah's problem as well as the positive aspects of the counselor's approach to beginning the interview, developing the interview, empathic understanding, and exploring of social behavior in terms of assertion, nonassertion, and aggressiveness. After reviewing the positive aspects, discuss ways

in which the interview may be improved.

Repeat and tape the interview for 5 minutes. Listen to the tape and then repeat the discussion for 10 minutes with the additional examination of the improvement shown in the second interview. Discuss possible reasons for each individual's role selection as the professor, nurse, counselor, or Sarah.

Sarah reveals a great deal in her stream of consciousness discussion. Unable to show her feelings, perhaps because she has been conditioned aversely (negative reinforcement) with a rebuff such as "Don't tell me your troubles" and/or annoyance (negative reinforcement) for imperfections or any show of weakness, she has become impelled toward "doing things" and doing them "right." Sarah seeks out situations in which she will become distraught; then she inflicts self-punishment because she is unable to assert herself with people—masochism. Her primary assertion is in a perfection orientation. Only rarely do her feelings spill forth in aggressive attacks on people. When she does let loose, she becomes very irritable to the point of seeking to belittle and rattle the individual at whom her wrath is directed—sadism. She turns from inward punishment in the form of headaches to outward punishment in the form of anger. A review of her learning experiences would support the inference that she has been reinforced for both kinds of behavior—masochism/nonassertiveness and sadism/aggressiveness.

A recent film, "Child's Play," based on a play by Robert Marasco, portrays with numerous shock effects the social relationship between aggressive and nonassertive behavior. A self-righteous teacher in a prep school for boys feels justified in destroying the career and sanity of a rival colleague. He despises the other teacher and handles this hostility with aggressive finesse. Subtly, but definitely, the aggressor imposes many environmental upheavals that are unmanageable by both the besieged teacher and the boys in the school. The nonassertive teacher is unable to withstand the onslaught of these upheavals. The boys are reinforced by the menacing teacher to perform sadistic (aggressive) acts on one another as well as on the masochistic teacher who blames himself for the school happenings. Child's play, indeed, with a note of horror!

The important lesson that may be taken from this drama of aggression and non-assertion is the fact that the two behaviors balance one another and are similar to sadism and masochism in this respect. A vicious cycle of malicious consequences perpetuate the behaviors. The self-reinforcing contingencies are established so that one act stimulates another act to follow it because of the self-satisfaction and other reinforcements that result. These contingencies become part of the process conditioning each boy to continue his sadistic and/or masochistic acts as well as the continuation of the group effort expended (group conditioning of actions).

Before the client is ready to alter his inappropriate behavior (aggression, non-assertion), he must recognize the what, when, and how of his behavior. Then he must want to do something about it. The interviewer's work is to help the client in both of these steps toward assertion.

Assertive training procedures

Preparation of the client for change. Interpersonal living is learned. Numerous authors have discussed the many forms of behavior resulting from this learning. Adler (1964), Alberti and Emmons (1970), Wolpe (1969), and Horney (1954) make similar distinctions in explaining social behavior. Adler writes of the "ruling attitude" and the "getting type," expressed in aggressive behavior, that are similar to Wolpe's "rough-shod, self-interest behavior" and Horney's description of persons "moving against others." All these labels for extreme egocentric (selfish) behavior are examples of a philosophy of "attack to conquer." In its extreme form this everything-for-self behavior becomes the psychopathic behavior of the individual who has not been reinforced to feel guilty or anxious about his behavior. He does his own thing no matter who gets hurt or how he must hurt them.

At the other extreme there is the non-assertive individual whose aggression is directed inward. He feels too much guilt and too much anxiety. He is the "side-stepping" type (Adler, 1964) who avoids problems. He "moves towards people" (Horney, 1954), often humbling himself to attain the reinforcement of other people's approval. He may be the handshaker, the backslapper, or the "yes" man who would prefer to avoid these behaviors but is keeping out of trouble in this way. In putting himself down he avoids the challenge of proving himself or, conversely, proves himself by allowing others to assert themselves, thus getting the thrill of "how wonderful am I to let him fulfill himself." He "moves away from people" (Horney, 1954) because he is fearful of the possible rejection and/or failure of social confrontation.

Assertion requires more than being free to affirm oneself. It also requires social interest (Adler, 1964). When the client has reviewed these differences among aggression, non-assertion, and assertion, he is better prepared to assume responsibility for investigating the details of his own behavior.

One more idea may promote the client's eagerness to attempt to do something about his inappropriate behavior—the concept of reciprocal inhibition. Wolpe (1969) explains that when an individual is able to assert himself, the resulting positive feelings of accomplishment are self-reinforcing. With the gradual conditioning of new habits of assertive responses, the individual is able to better curb his anxiety feelings. When the client does something new by word or act to move himself forward (positive reinforcement), he inhibits former unsatisfying behavior (negative reinforcement). The new habit serves a protective function through release of feelings and, in the process, reduces anxiety. An individual who suddenly thinks of something humorous while embroiled in a blazing argument finds that the fire of the verbal fight begins to cool. This, too, is reciprocal inhibition.

Potential difficulties. Before the interviewer can help the client become more assertive, he must identify some client characteristics that result in difficulties for the client. There are perplexing characteristics that the interviewer may encounter—generalized inappropriate behavior, fear of aggression, and the proper units for behavior change.

Generalized inappropriate behavior is a combination of behavioral characteristics that must be separated by a more complex approach than can be accomplished in the usual counseling situation. Thus a more experienced professional may be needed on a consultative basis or the client should be referred to someone else who is able to handle the more complicated problems associated with the general patterns of frustration and failure.

In contrast to the widespread theme of defeating behavior, the more selected the conditions under which nonassertion or aggression occurs, the simpler the solution for behavioral change. The individual who is nonassertive when he is with individuals he considers more learned than he or who becomes extremely (and sometimes obnoxiously) forceful when debating a point with a certain person has a greater possibility for rapid success in changing to an assertive approach than the person who exhibits these inappropriate behaviors in most social situations.

Another kind of problem originates with the individual who evidences fearful responses, even to the point of being phobic (irrational, pervasive fear) about self-assertion. The anxiety-stimulating events are often extensive and the individual may be concerned with speaking to a group of people, asking questions, or any form of social situation. Assertion either becomes the trademark for aggressive behavior that horrifies the individual or an overwhelming form of one-upmanship that the individual considers undesirable. Extreme caution must be used not to break down the barricades of coping strategies too suddenly. Whether or not this person should be referred to a therapist rather than a counselor must be one of the considerations before the interview process is begun.

Finally, the interviewer must plan precisely in order to decide on comfortable units for the assertive training process. These units of approximations toward the terminal behavior of assertion by choice should have a high probability of success. The anguish that the client has experienced from prior attempts and failures must be avoided. Sometimes failure and setback may follow when the client misinterprets a cue or is too clumsy in his newly sprouting behavior. Then the interviewer must step in to help the client understand and regain his confidence.

Behavior shaping. The process of desensitization is an important element in the conditioning of assertive behavior. In desensitization a hierarchy of anxiety-producing stimuli is drafted and the client is presented with these stimuli (for example, gradual and closer exposure to feared snakes)

and positively reinforced at each unit of success accomplished. This is shaping.

In the shaping process the terminal behavior is broken down into small units or single responses. These are arranged in a hierarchy of activities of increasing difficulty. Each step requires a little more of the individual. After the individual makes a successful attempt the interviewer positively reinforces him either verbally or nonverbally; then the next step is assigned. Each new unit should be simple enough that the client may continue without too much stress.

The process of shaping is encouraged when the interviewer models both the inappropriate behavior the client exhibits and the appropriate behavior to be achieved. The client observes and is encouraged to assess the interviewer's modeling. Then the interviewer may once again model while the client takes the role of the other person involved in a social situation. This process may be repeated until both interviewer and client are assured that the difference between nonassertive or aggressive behavior and assertive behavior is understood. An additional procedure may be added during which the client role-plays himself, first with the inappropriate and then with the appropriate responses. The use of video tape or an audio tape increases the feedback for the client. For some clients, practice in front of a full-length mirror helps them note nonverbal cues.

After the client examines his own behavior as well as the interviewer's modeling behavior, he begins progressive steps of changing his behavior toward the goal of assertion. During this process the interviewer is actively engaged in modeling and providing feedback to the client about his own role-playing by means of a discussion that is enriched with the video and/or audio tape.

How long the process takes depends on the client's degree of motivation, what behaviors are already possible for him (his behavioral repertoire), the success at each step toward the goal, how carefully the interviewer has recorded the client's progress, and the reinforcements he receives from the interviewer and from the significant people in his life space. In addition, the decision about when to progress from practice in the security of the interviewer's office to a real social situation calls for a joint effort of brainwork and agreement by both the client and the interviewer. This may require only one visit or a series of visits to the interviewer's office.

The client who launches upon trying out his new response patterns is more likely to succeed if his assertive responses have become automatic. The client should report to the interviewer his attempts to behave more assertively rather than with his former inappropriate nonassertion or aggressiveness, and together they should review the degree of success experienced. This may be accomplished through discussion or by means of role-playing.

The final step in learning assertive responses is the acquisition of a generalized behavior pattern that permits the client to expand his assertive responses to situations other than those practiced within the confines of the interviewer's office. Eventually the client realizes that the process of living entails persistent testing of his own behavior in a multitude of situations and the ability to be self-reinforcing.

Changed self-concept

The terminal behavior to which the client is striving is assertive behavior that is satisfying. Thus the client who becomes sufficiently assured of his assertive ability can discriminate the situations when he will assert himself. Since he does not have to hide or to prove himself constantly, he can look more readily to the needs of others. He is able to wait for a favorable time for assertion when he will not be upsetting the more sensitive person or when too many "bad vibes" will make assertion fruitless.

Furthermore, he is able to admit he is wrong when he makes a mistake and not compound the mistake with "assertion" colored red with the fury of aggression. Indeed, he will know when compromise is the better path to follow rather than stubborn resistance to have only his way.

To assert or not to assert depends on the insistent essential of social interest (Adler, 1964). Social interest demands that solutions be useful not only to oneself but also to others. Thus the individuals who have adopted social interest as a way of life "see with the eyes of another, hear with the ears of another, feel with the heart of another" (Ansbacher and Ansbacher, 1964), yet are aware that these are borrowed on a tempo-

rary basis. Empathy and social interest have much in common.

With the ability to assert himself—by choice—the client comes closer to following the Talmudic saying, "If I am not for myself, who will be for me? But, if I am for myself alone, what am I?"

EXPERIENTIAL INTERVIEW

Sometimes it is difficult to draw sharp distinctions between the information-oriented, behavior modification, and experiential interviews. The information-labeled interviews may be concerned with more than objective factual data and may look for findings about attitudes, values, feelings, hopes, plans, and self-descriptions. The behavior modification interview may also take into account more than changing inappropriate behavior. Behavior change is more efficiently accomplished in an atmosphere of empathic understanding, genuineness, and nonpossessive warmth (Truax and Carkhuff, 1967).

Perhaps the more significant differentiating features are the kind of sharing and understanding that enter into the experiential interview and the aim toward some degree of personal growth for the client and for the interviewer. These features may also occur in the information interviews but are then side effects rather than primary goals.

The experiential interview stresses the human relationship between two people being together in a moment of time. The immediate situation and the sharing of thoughts in an acceptant atmosphere are essential attributes. The experiential orientation erases a hierarchy of "better than" and "worse than" and thus omits judgmental reactions. The helping-knowledgeable person and the client join in their quest for personal growth and in becoming more fulfilled individuals. How many interviews and how frequently the interviews occur becomes a mutual decision.

Experiential means just what the word implies, experiencing with the client. This kind of interview becomes a microunit of life in which two people, the interviewer and the client, share feelings, understandings, and behavior—a small unit of life experience in which the individuals involved share themselves. The experiential interview is the type of interview that is the focus of this book.

Examples 5-4 to 5-7 compare the information-seeking, the information-sharing, the behavior modification, and the experiential interviews.

Example 5-4
INFORMATION-SEEKING INTERVIEW

MS. H.: *(Knocks at the door, then bends down to pick up the briefcase she has put at her feet.)*

HELEN: *(Opens the door slightly.)* Yes?

MS. H.: I am from the University of Wannuga. We are conducting a survey for the Institute for Policy Studies on current opinions toward women's liberation groups. See, here are my credentials.

HELEN: *(Opens the door a little further.)* I'm really very busy today; have to finish something before I leave for an appointment.

MS. H.: Just 10 minutes. The questions won't take more than 10 minutes to answer. We have selected five houses on this block and since yours is in this chosen sample, it would help considerably if you would give me 10 minutes of your time.

HELEN: *(Opens the door wider and Ms. H. enters.)*

MS. H.: Thank you very much. I appreciate your assistance. You do have an attractive house. May I sit here?

Ms. H. removes the questionnaire from her briefcase and proceeds to complete the background information about the educational level, financial group, occupational level, and age group that describes Helen. Then Ms. H. asks the questions on the questionnaire, stopping to chat briefly when Helen asks for some clarification about the purpose of the study.

Example 5-5
INFORMATION-SHARING INTERVIEW

TOM: *(Enters the counselor's office at the college and sits down in the chair opposite Ms. M.)*

MS. M.: *(Looks up from her desk and smiles.)* Hello, Tom. Nice to see you again.

TOM: *(Stretching out his legs.)* I've been thinking about our conversation last week and

looked into *Barron's Guide to Colleges*. Need to talk to you some more about some of my ideas.

MS. M.: *(Silent.)*

TOM: Last week I was tossing around whether I should take off 1 year after graduation, then go on to graduate school, or start grad school right away. Want to work. Just work. Doesn't matter what I do. Want to be around some people I like—people who dig me, too. Yet my parents are on my back about grad school.

MS. M.: Work or school? Difficult decision to make. So many things going for either decision.

TOM: Hmmm, that's it! When we talked about it last week, you suggested *Barron's* to see if any of the colleges turned me on. I did. Saw a few colleges—offer the program I want, in the geography I want, people I want, near enough. I could get in, but. . . . *(Pulls up his legs and sits forward. With his hands clasped between his knees, he looks over Ms. M.'s shoulder.)*

MS. M.: *(Silent.)*

TOM: Gotta get away from education bullshit for a year.

MS. M.: Have you any ideas of where you would go to work for the year?

TOM: New York. Like to be around New York. Don't know where I would find work or a place. Some people said they'd come to live with me.

MS. M.: I have some information about job opportunities and apartments in New York. Would you like to look the lists over during the next week and then we can get together to talk about your decision for next year?

TOM: Thanks. Give me a start. I'll keep *Barron's*, also, and see you at 2 next week.

Example 5-6
BEHAVIOR MODIFICATION INTERVIEW

With stooped shoulders, the man and the woman slowly walk into the psychologist's office. The woman holds a crumpled tissue in her right hand. The psychologist, Dr. E., rises and walks over to Mr. and Ms. R.

DR. E.: Hello, please sit down over here. *(He motions to two black leather seats that face a third black leather seat.)*

MR. R.: *(Sliding down in the seat.)* Well! This is a comfortable chair.

MS. R.: *(Looking at her husband.)* Oh, Tom, let's get down to it. This uncertainty is . . . Dr. E., is he— is Buddy really not— not . . . .

MR. R.: Not normal—not normal. That's what we want to know. Is our child normal?

DR. E.: Dr. M. examined Buddy, and he believes Buddy has a good chance to live a satisfactory life. Buddy is a jolly 5-year-old who can do much more than he does. I administered some psychological tests and find that Buddy functions at the intellectual level of a 2-year-old, but both Dr. M. and I believe he is able to do much more.

MS. R.: But Doctor, it just can't be—can't be. What did we do to—Doctor, you're the fourth doctor we've seen. Just confuses us—ready to give up. This doctor shopping won't do.

DR. E.: That's fine. You are ready to really help Buddy now, to help Buddy learn a great deal.

MS. R.: He looks so funny—eyes so far apart, nose so small, chin almost not there, ears so low. How can he live a happy life? He's so unmanageable, bangs his head against the wall. Have to take care of him, dress him, feed him.

MR. R.: Come now, Doris, you don't even give the kid a chance to do anything on his own.

MS. R.: What do you know? You don't have to watch the mothers of the other children look at me with pity when I have to carry Buddy up and down the steps. You don't. . . .

MR. R.: Shut up, Doris. I don't want to hear you any more.

MS. R.: *(Dabs at her eyes with the crumpled tissue.)*

DR. E.: Buddy can do much more. It all depends on reinforcement, systematic rewards for appropriate behavior.

MR. R.: What do you mean—reinforcements? What more can Buddy do?

DR. E.: Look at it this way. Buddy learns just as anyone else would but more slowly. Behavior is caused; it can be changed by what happens after the behavior, in other words, the consequence of the behavior.

MR. R.: Yes. So, how do we do it?

DR. E.: It's much easier than you think. Both of you can help.

MS. R.: Doctor, I don't know—don't. . . .

MR. R.: Doris, hear him out.

DR. E.: There are three important parts to changing behavior. First, you have to decide on the behavior you want to change. This is your target behavior, your behavioral goal. Pinpoint this behavior. What behavior bothers you most?

MS. R.: Head-banging, that's it. Can't stand it; I yell at him; I hold him in my arms; I talk to him, but he does it more. When I don't let him do something, then that's the time he does it.

DR. E.: Fine. Now you have given step two, what follows the behavior, and step three, what happened before. Target behavior— head-banging; reinforcement—yell, hold, talk equals attention, and you say that what happens before is stopping him from doing something. There it is, that's your formula for change.

MR. R.: Sounds easy. Is it that simple?

DR. E.: Simple, yes, but it may be made difficult because you need to be consistent, particularly at the beginning of the learning. You must decide on reinforcers to extinguish the behavior. For Buddy, ignoring the head-banging would be a form of negative reinforcement. Pay attention when Buddy does not bang his head, not when he does.

The interview continues with Dr. E. practicing behavior modification with Ms. R. Whenever Ms. R. becomes upset, he ignores her. When she or Mr. R. makes a statement that demonstrates willingness to do something positive about Buddy's behavior, Dr. E. pays attention by responding.

Example 5-7
EXPERIENTIAL INTERVIEW

Dr. P. is seated at his desk when he hears a gentle knock on his office door. He rises from his chair and goes to the door, opening it slowly. Before him stands a small-framed young woman with long, blonde hair. Dr. P. notes the frown on her face and the down-turned mouth.

DR. P.: Hi! I'm Dr. P., come in. *(He points to his left where there are two chairs, one straight-backed and the other upholstered in brown leather.)* Please sit down.

TERRY: *(Sits down on the straight-back chair with her hands loosely lying in her lap and her feet pushed under the chair seat.)* I phoned yesterday. Told you about myself. A little bit. I'm Terry.

DR. P.: Yes, I remember. You said that Margie told you about me.

TERRY: *(Staring at her hands and silent.)*

DR. P.: Perhaps it would help if you would tell me a little more about how you feel.

TERRY: *(In a low voice.)* As I told you over the phone, I'm pregnant. Just found out definitely from the doctor 2 days before I phoned you. Can't tell my instant-liberal, middle-class, hung-up parents. Thought I could. Have to know what to do . . . talk to someone. I want an abortion, but don't have the money . . . don't know where to go. Haven't told Marty about it. Don't want him to know. So, I feel alone, pushed in, lousy.

DR. P.: Hmmmm. You feel confused, need someone to listen, someone to understand and help. I know that feeling.

TERRY: Yes, help. I spoke to Margie and she said if she were pregnant, she would have the baby. She's a year older, 16, don't think that makes her readier. She belongs to the high school women's lib group. I don't. I'm too young to have a baby. Want to finish school. How would I face my parents . . . or my straight friends?

DR. P.: *(Silent.)*

TERRY: Marty is a good person, a beautiful person. He knows where he's at. Someday we may get married. Maybe not—too young to know, too much still to do. He's 18 and in his first year of college. He's the only one. We've been total for about 6 months. Don't you think an abortion is the only answer?

DR. P.: I wish I could give you a simple "yes" or "no" to your question. It might or it might not help you. But the answer is yours. I will help you try to find the answer. Some facts might help. How far along in pregnancy are you?

TERRY: Five or six weeks, I think. That's what the doctor thought. Wow, was that something, telling him I'm Mrs. G.

DR. P.: You do seem to have some strong feelings about being pregnant. It might help to air those feelings.

TERRY: You know, that's the odd part. Feel angry at Marty, yet like the idea. Makes me feel a little closer to him. Feel that I would like to have the baby . . . to sort of play with it. Not ready for that mother game. Feel mad at my parents because I don't trust them to let me talk about this. Feel guilty about not being careful and maybe fearful about an abortion. Feel as if I'm going back and forth. *(Sways slowly back and forth and her hands become fists in her lap.)*

DR. P.: Yes, I see, I see.

TERRY: *(Silent.)*

DR. P.: I feel that you're saying—you seem to be telling me that you're both glad and mad at what has happened, that you wished that your parents and Marty would know about how you feel so that they might share the burden.

TERRY: *(Looks at Dr. P. and stretches her hand out toward him.)* That's it . . . mmmm . . . that's it. It seems as if I'm left alone when I need someone most. Marty should know about it without me telling him. My parents should offer their help without me asking for it. Feel good about it. Feel bad about it. If only I could wish it away, all away.

Exercise 5-4
ROLE-PLAYING VARIETIES IN INTERVIEWING

An effective method to discover the differences in the atmosphere and the different approaches described in the four previous examples is to role-play the different situations.

Get together a group of five trainees. Three members of the group will be the observers and two members will be the role-players. The role-players should continue the interviews from the point the actual conversation stopped in the examples. The three observers should take notes about different aspects of the role-playing situation. One member should note verbal cues, another member notes nonverbal cues, and a third member should observe the entire situation noting all cues.

The two role-players decide who will be the interviewer and who will be the client and then conduct an interview for 5 minutes. Stop the role-playing and discuss the observational cues that are recorded. Discuss the kind of interview that is developing—information getting, information sharing, behavior modification, or experiential. Suggestions should be made by all members of the group as to how the interview is effective and how it might be even more effective.

Role-players resume their roles but this time the interview incorporates the suggestions from the discussion. Observers once again record the verbal and nonverbal cues. After 5 minutes stop the role-playing and again discuss the recorded observations.

CHAPTER SIX
THE THREE PHASES OF THE INTERVIEW STRUCTURE

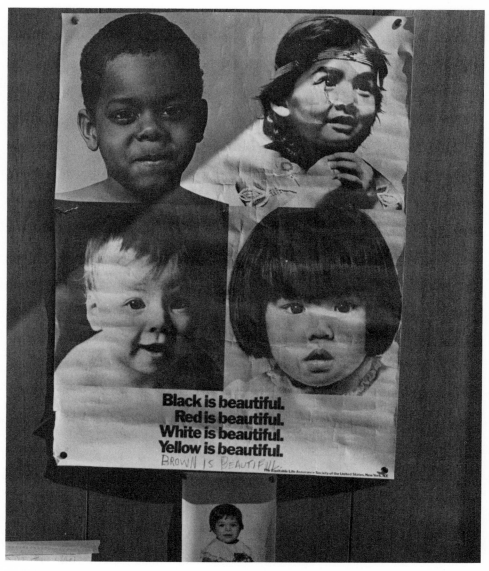

Acceptance: all people are worthy of respect. (Photo by Irwin Nash; from Wagner, N. N., and Haug, M. J., editors: Chicanos: social and psychological perspectives, St. Louis, 1971, The C. V. Mosby Co.)

Every single interview with a client as
well as the conduct of a series of interviews
with the same client follows the same structure. This structure involves the opening
or beginning phase, the developing or exploring phase, and the ending or terminating
phase. These three phases overlap.

OPENING OR BEGINNING PHASE
Establishing the relationship and roles of interviewer and client

The interview begins when the client first
makes contact by telephone or in person
with the interviewer or a secondary person
who relays the message to the interviewer.
The client's feelings about this initial contact will have much impact on what happens
later in the interview process. First impressions do count, particularly with the troubled person.

A client probably has preconceived ideas
of what an interviewer's role should be. The
client may think of the interviewer as a
question-answerer, as someone who will tell
the client what to do, how to do it, and
when to complete the "doing." On the other
hand, the client may be searching for someone who will silently listen to his complaints
and preferably agree with his solutions. It is
important that the client's notions about the
interviewer be part of the opening discussions.

What the interviewer does depends on the
client. No roles or rules adequately specify
the interviewer's behavior. The way in
which the interviewer responds to the client
must be in terms of what would most benefit
the client. The interviewer is primarily a
person committed to *help* the client—a
person who is not hung-up on a particular
assumed role because of his own unfulfilled
needs.

Only one characteristic is similar for all
clients. All clients *voluntarily come* or *are
brought* to the interviewer for some assistance, *for help.* Only one characteristic is
similar for all interviewers. All interviewers
must be functioning at a higher level than
the client. The interviewer must be the
knowing person who can and will give
help.

Stating the purpose of the interview

One of the first tasks in the beginning
phase is to state the purpose of the interview.

The similarities or differences among the
expressed purposes of the client, the interviewer, and other interested parties provide
the interviewer with some understanding of
the degree of pressures and distortion involved in the client's problems.

From the client's viewpoint

From the client's viewpoint the purpose
of the interview depends on whether the
client has sought help or was sent or brought
for help. The self-referred client is more
likely to ease into the client role, a seeker of
help. The purpose of the client's seeking
help is important even if the interviewer
or other sources see another picture. Establishing an empathic relationship requires
the interviewer to see events through the
client's eyes.

From the interviewer's viewpoint

From the interviewer's viewpoint the
client's expressed purpose for the meeting
may appear to be only a cover for a more
complex bewilderment that is being concealed by surface concerns. The client may
feel safe in relating only a small part of his
concerns at this point in the interview relationship. In the initial phase the client may
be unaware of the connection of several
problems.

The takeoff for the interview must be
where the client states the beginning should
be. The skilled and experienced interviewer
will assist the client in a greater degree of
self-exploration. The interviewer's job is to
help the client break down the walls of the
tunnel through which the client views his
world. The client must be prepared to
expand and delve.

From the viewpoints of other interested parties

From the viewpoints of other interested
parties the client may appear to have a
more or less serious problem than he expresses. Often the degree of seriousness is
judged by others in terms of the acting-out
behavior of the client. The different tolerance levels of the people around the client
influence which behavior is considered a
problem. A close relative may be frightened
by the argumentative behavior of the client
or annoyed by the unceasing complaints.

Yet the withdrawn client who makes no waves within the household may be ignored.

• • •

Example 6-1 compares the purposes of an interview from the client's, the interviewer's, and the client's mother's viewpoints.

Each individual in Example 6-1 has only one view of the client's problem. The purpose of the interview as the client sees it is to help him get out of a school situation that is presently overwhelming for him. This is the manifest purpose for the interview—the top layer of the client's problem. The interviewer is searching for more physical and psychological information about the client. The interviewer is just scratching the surface of the client's problem. The mother describes her purpose for the interview in terms of the annoyance value of her son's behavior. Her purpose is to stop his acting-out behavior and to make him toe the line. The mother appears to have no understanding of the seriousness of the client's problem nor is she aware of her son's drug usage. Even though the purposes of the client, the interviewer, and the mother differ, together they provide a more complete picture of what factors might be entering into the interview situation.

Understanding the expectations for the outcome of the interview

Outcome expectations are related to the expressed purposes but should extend beyond those initially stated. The 15-year-old boy wants out of school according to his expressed purpose. Yet, as the interviewer notes, this client's problems are interwoven with maternal pressures and parental misunderstanding. The communication barriers between parent and son had to be removed. Therefore the initial purpose was just that, a beginning.

Expectations for the outcome are related to one other item in addition to the expressed purposes of the client, interviewer, and other interested individuals. This item is the time involved. The amount of time involved influences when the outcome can be expected and the procedure to be followed in attaining the outcome. One simple statement may be made no matter how much time is involved—the interviewer must make clear the amount of time available for the interview but should not stretch out the interview just to fill time. In addition, the interviewer must be cautious in making the available time known to the client so that the client will not feel pressured to hurry or too frustrated to continue.

The novice interviewer is more likely to find it difficult to use the allotted time for

Example 6-1
VIEWPOINTS ABOUT PURPOSE OF INTERVIEW

CLIENT'S PURPOSE	INTERVIEWER'S PURPOSE	MOTHER'S PURPOSE
Fifteen-year-old boy speaking to interviewer: "I've been smoking pot now for 2 years. Fine. Like it. Tried firing*—scared. Scared because don't feel so good. Don't want a monkey on my back. My parents and school are heavy enough. Can you get me out of school?"	Interviewer thinking to himself: "This 15-year-old needs medical care. Looks thin and worn out. Must lead toward a medical exam soon. Wonder whether he fired only one time. Need to find out more about family relationships and school."	Client's mother telephoned the interviewer: "He's a tyrant at home. Yells, uses foul language, threatens to hurt his younger brother. I don't know what's gotten into him. He either locks himself into his room, mopes around the house, or runs through the house shutting all the windows and locking all the doors. I think what he needs is discipline. Do you know of a good military school where we can send him?"

*Firing refers to injection of the drug into vein. In this case the drug used was heroin.

the interview. The interviewer may encourage more talk because of guilt feelings or feelings of inadequacy. The interviewer may feel that he has not accomplished enough, so he pushes some more. The fee must be earned, so he continues professional services until the completion of the 45- or 50-minute "hour."

Whatever the motives of the interviewer, the client will not be helped very much after the stopping point has been reached. The interview should be stopped at the moment when both the client and the interviewer agree that all has been said. The interviewer is more effective if, when the client sends out stop signals, the interviewer comments, "Would you like to stop now? I notice you looking at your watch." The words may differ, but the sentiment should be disclosed.

Since the time factor is part of the interview process, the time framework dictates what must be accomplished. The suicidal client relating his intentions and the methods to be used is in a crisis interview. The telephone becomes the vehicle to *immediately* give the client reason to live or, at any rate, a reason not to die at that moment. Thus the time involved in a crisis interview pushes together the opening, developing, and closing phases.

In other interview situations an immediate goal may be satisfied in a single interview or the interview may become the first of a series of interviews. The client with the interviewer sets the short-term outcome for each interview. They also plan the long-term outcome for a series of interviews. The short-term and the long-term outcomes are flexible and serve as guides for the expected outcomes between the interviews. Example 6-2 relates time to outcome.

Example 6-2
TIME AND INTERVIEW OUTCOME

Enid is 10 years old. The possible diagnosis is autism (severe emotional disturbance associated with withdrawal and language disturbance; capable of object relationships but little or no interpersonal relationships).

First interview: short-term outcome
(**some response to interviewer**)

Enid sits in the chair near the wall. She presses her head and shoulder to the wall. Her legs are close together, her back is straight, and her arms are folded tightly on her chest. Slowly she turns her head forward with fixed eyes blankly focused straight ahead of her.

The interviewer, Ms. S., had smiled whenever she had passed Enid during the past week since Enid had arrived at Dencrest Center. Ms. S. gets up from her chair, moves closer to Enid, and stops in front of her.

Enid looks up, unfolds her arms, and places one hand on the chair next to her. Ms. S. sits down and after a few minutes places one hand on the edge of Enid's chair. Enid and Ms. S. sit silently. Enid remains sitting rigidly upright with her feet on the floor. She does not respond to Ms. S.'s hand on the edge of her chair. After about 10 minutes, Ms. S. stands before Enid, smiles, and says, "I'll see you again tomorrow, Enid. Thanks for coming."

Second interview: short-term outcome
(**some verbal response to interviewer**)

Enid is brought to Ms. S.'s office by one of the health aides. Ms. S. is waiting for her, seated in the same chair as on the previous day. Ms. S. places her hand on the edge of Enid's seat and Enid rigidly walks over to the chair and sits down. Ms. S. observes that there is a difference in Enid's movements today. Enid does not push her shoulder up to the wall. Instead, she places one hand on the side of her thigh. Her arm and hand near the wall are held tightly against her body. Enid's hand rests near but not on Ms. S.'s hand. Ms. S. turns slightly so that she faces Enid at an angle.

MS. S.: *(Smiles.)* I am near you, Enid.
ENID: *(Quickly looks at Ms. S.)*
MS. S.: How are you, Enid?
ENID: *(Again looks at Ms. S.)*
MS. S.: How are you, Enid?
ENID: *(Mumbles.)* Hurting.

Long-term outcome
- Establish trust in interpersonal interactions with Ms. S.
- Encourage conversation, first with Ms. S., then with staff, and later with peers.
- Begin education at center with further possibility of entering school outside of center while living at center.
- Further goals to be determined depending on progress.

The goals that Ms. S. proposes are based on the time factor as well as Enid's problems. Simple and concrete goals with favorable possibilities for accomplishment are planned. Since relating to people is difficult for Enid, Ms. S. includes brief exposure to the short-term outcomes of interpersonal contacts. Each new exposure is arranged to begin where the previous interview left off. Note that in the second interview Ms. S. sits in the same place with her hand held in the same way as at the end of the first interview.

Ms. S. is aware that the fuller development of trust and a relationship will take much longer to achieve than the simple responses of Enid's placing a hand on the chair next to her or mumbling something in response to Ms. S.'s question. Yet each change toward the final goal is a step forward and, for Enid, an achievement never before completed.

The time allotted to each interview is brief. The frequency of the interviews allows for small units of behavioral outcome. In another situation the arrangements would have to fit the time schedule and the client's needs.

Avoiding interview pitfalls

The behaviors identified in this section are called pitfalls because they represent unsuspected difficulties that may trap the interviewer or the client and prevent the client from fulfilling his goals.

Steering

Two forms that steering may take are the Greenspoon effect and the prompting effect. Both of these steering devices bias the interview and disrupt the client's direction of thought.

Greenspoon effect. The Greenspoon effect is named after J. Greenspoon, who first noted the dramatic effect of certain sounds (mmm-hmmmm and huh-uh) on a client's verbal behavior (Greenspoon, 1955). Both the client and the interviewer may be unaware of the source of the bias or the extent of the influence of these sounds on the client's responses. However, unless the interviewer is aware of particular speech mannerisms, he may steer the client's responses away from the main issues of concern to the client.

The communication distortion that may result from the Greenspoon effect will be directly related to the suggestibility of the client. The client may not "tell it like it is" but tell it like he is being reinforced to tell it.

Prompting effect. Further contamination of the interview may arise from the tendency of some interviewers to introduce their own ideas into the conversation. The interviewer may comment on a client's answers, may suggest answers, or may use nonverbal prompting. The following samples of interviewer prompts demonstrate the different forms that interviewer prompting may take.

In the first sample the interviewer (Mr. P.) prompts the client by commenting on the client's answers.

MR. P.: Yes, I see how that might upset you. How did you handle it afterwards?

GREG: I was on a total trip. Felt as if all stops were gone. Said things to my father that dug deep. Told him what a jerk weakling he is, screamed at him. Why doesn't he stand up to that b-b-b . . . . Oh, what's the use? Done—my mother . . . . *(Voice trails off.)*

MR. P.: You seem to think your mother is the cause of all your problems and your father's, too.

GREG: Mother? Hmmm, may be. Is that what you think? Maybe you're right. My mother—yes—didn't realize that. Hah! The life-giver—the living-taker. Damn her, you're right. Yes, yes, right!

Sometimes the interviewer prompts by suggesting answers, as may be noted in the following sample.

LESLIE: Just have to find out. Should I go along with Ches? All my friends didn't tear their thoughts to decide on what they should do. Can't decide! What's marriage anyway? Who needs it?

DR. G.: You're really working hard at this. Do you feel as if both establishment pressures and less conventional aspirations are gnawing at your decision?

LESLIE: Yes. That's true. When Ches and I talk about us, the logic is on his side. But when I'm home with my parents . . . I wonder how they would feel. Really don't want to hurt them. Don't know what to do or say. What should I tell Ches? Should I tell my parents?

DR. G.: *(Silent.)*

LESLIE: Tell me, Dr. G., what do you think?

DR. G.: You should discuss this with your parents. Tell them that you and Ches are not ready for marriage but have a great need to be together. Tell them—ask them . . . . (*Dr. G. speaks at length in answer to Leslie's question.*)

Perhaps the most subtle form of prompting is provided by nonverbal cues.

MS. I.: Hello, Bob. (*Smiling.*) Come in, come in. Sit down in that comfortable chair near me.

BOB: Ms. I., I can't go on with the weight of this. Tina and I want to get married. Remember I told you about her a few months ago? We do want to marry. Have plans and wedding arrangements. One thing I didn't tell you—Tina has the sickle cell anemia trait. And me. Wow! I have it. Disease right down the line. Ms. I., are you looking that way because you are thinking as our parents do, no marriage?

MS. I.: Frowning, is that what I was doing? Yes, guess I was thinking that you would be starting marriage with a couple of strikes against you.

BOB: Tina is a nurse so she knows about the possibilities of our children inheriting sickle cell anemia. We both know that whether our children would have the trait or the disease depends on how the genetic dice roll. But . . . .

MS. I.: Yes, you do have a heavy decision to make. (*Ms. I.'s frown is again apparent. Ms. I. begins to tap on the desk with one of her fingers.*)

BOB: Have I upset you? I can sense your displeasure. This is not easy. In fact, now I'm more indecisive.

Blatant or subtle prompts can and do steer the client into the interviewer's mold of thinking.

Allness

Some of the steering described previously may be related to the need of some interviewers to be one-up on their clients. Some interviewers revel in the client's effort to endow the interviewer with magical powers.

The interviewer's need to be all-knowing, all-powerful, and all-capable may result in a power struggle that the client often loses, leaving him feeling more helpless and inferi-

or. At other times the interviewer may suggest symptoms or illness to the impressionable client. In an effort to assure the client that he is knowledgeable the interviewer may offer more information and an untested diagnosis. Thus symptoms of a disorder are produced and sustained by the interviewer. Symptoms that originate in this fashion are called iatrogenic.

Insistence on the interviewer's infallibility may lead to the client's overdependence on the interviewer and the implication that no other individuals understand the client as well as the present interviewer. Furthermore, the power-driven interviewer may exaggerate findings and even argue with and badger the client.

Omnipotent (all-powerful) needs may lead the interviewer to use words and concepts over the patient's head. Then the interviewer may enjoy the client's failure when the "Great Interviewer's" words have not been followed. This sounds harsh, but there are interviewers who, because of their own unfulfilled needs for recognition, unwittingly perhaps express themselves best in the presence of the client.

The most serious consequence may be if the interviewer enjoys or is embarrassed by the client's discomfiture even to the point of laughing. This form of cruelty to clients happens with novice interviewers who are uncertain about procedures and too eager to be impressive.

Mistakes in approaches to clients are bound to occur. Inconsistencies, inaccuracies, and exaggerations are not unusual. The problem goes much deeper, however. The real problem lies with the interviewer who makes no effort to correct errors. The more serious destructiveness stems from the inability of the interviewer to admit errors because of a profound need for prestige and power. The beginning interviewer is able to alter such an allness course through self-understanding before his reputation becomes more important than the client's welfare.

Professionalitis

Professionalitis is closely allied to allness. Because of feelings of inferiority or insecurity, the interviewer may hide behind complicated explanations, high-sounding jargon, and professional cliches. The tone of voice as well as the manner of speaking reflects the "professionalitis touch."

There is a difference between the professional interviewer who is a helping person and the interviewer who exhibits professionalitis. The professional interviewer is knowledgeable, skilled in interpersonal procedures, trained, and functions at a more life-satisfying level than the client. The professional shares understandings, while respecting the worth and dignity of the client. The interviewer who exhibits professionalitis feels impressive and tries to impress others with the authority of his superiority, his knowledge, his skills, and his training. Professionalitis requires that the client be barricaded from the inner sanctum of the professional mystique.

Examples 6-3 and 6-4 demonstrate different client problems and different interviewer approaches. Which of the two interviewers evidences professionalitis?

Example 6-3
MS. M.: PROFESSIONAL OR PROFESSIONALITIS?

MS. M.: *(Looks up from her desk as Ruth enters.)*

RUTH: Mr. G. sent me here. Thought I needed some help.

MS. M.: What seems to be the problem?

RUTH: Dunno. Mr. G. says I've been falling asleep in class. Not true, really. Only maybe once in awhile when I stay up late to study.

MS. M.: Hmmmm. So you think Mr. G. is analyzing the situation incorrectly? Or perhaps you think he's saying that you're malingering—you know, pretending to be asleep so you can get out of answering questions.

RUTH: Huh? What do you mean? When I sleep, I sleep. Sometimes everything gets a little fuzzy.

MS. M.: The idea of your being a malingerer seems to disturb you. Do you think your need to evidence your superiority over your brother might have anything to do with it? Sibling rivalry is not unusual.

Example 6-4
DR. Y.: PROFESSIONAL OR PROFESSIONALITIS?

Dr. Y. is administering the Draw-A-Person Test to Ned, 12 years old. The room in which the test is being administered is small and the walls are white cinder blocks. Ned comes into the room and looks around at the walls.

DR. Y.: Bare walls, aren't they? Hello, Ned. Please sit down.

NED: *(Sits down on the chair at the small desk.)*

DR. Y.: I would like you to draw something for me. Here, please draw a person on this sheet of paper.

NED: *(Picks up the pencil and rapidly draws a figure of what seems to be a girl with long hair and a dress.)*

DR. Y.: That's fine, Ned. Do you feel comfortable in that chair? It may be somewhat small for you. If you would prefer it, you might sit over here at my desk. More room.

NED: *(Looks at Dr. Y. and smiles. Rises and walks to Dr. Y.'s desk and slides into the upholstered desk chair.)* Like this chair. Yes, better now.

• • •

Ms. M. and Dr. Y. both appear to be knowledgeable. One of them, however, abides by the principle, "If you have it, flaunt it." Who does this, Ms. M. or Dr. Y.?

Pinch-hitting

Pinch-hitting is a trait that most humans develop to some extent. The mother who basks in the glory of her "son, the doctor," is pinch-hitting. The father who glories in the football prowess of his son is pinch-hitting. Another term for pinch-hitting is "vicarious experience." Vicarious experience refers to the satisfaction of one's own needs through substitution and imagined participation in the behavior and accomplishments of others. Vicarious living through others is not the same as being proud of another's accomplishments. Substituting someone else's experiences for one's own participation is usually accompanied by feelings of inadequacy and inferiority. Thus pinch-hitting is usually a cover-up for distress. The enveloping, yet hidden, thought of the nonproducing individual is, "If only I could do that."

Pinch-hitting by the interviewer does not promote personal growth for the client. Frequently the problems and needs of the interviewer creep into the interview. For example:

- Because of the interviewer's dependency needs, the client's dependency may be encouraged. The interviewer may treat the client as a helpless person who needs "parenting," thus decreasing the client's possibilities for self-growth. In giving too much concern, care, and help, the interviewer may feel almost as if these behaviors were offered to him.
- The sexually meek interviewer may revel in his client's sexual pursuits.
- The interviewer who is unable to combat the infringements by his parents or wife on his own interests and directions may encourage the client, openly or by a nod and a smile, to talk back or to act out aggressive behavior.

Living through the client's life is most unsatisfactory for both the client and the interviewer and must be recognized as one of the barriers to facilitative helping.

Outside/inside view

Another subtle cover-up for unpleasant feelings and unmet needs may take the form of either taking only an outside view of events, making an all-out effort to be objective (overobjective), or taking only an inside view of events, becoming sentimental and overinvolved emotionally (oversubjective).

Overobjectivity and oversubjectivity are two widely divergent approaches. Overobjectivity inclines toward coldness and insensitivity. To get the facts without being aware of feelings may result in a robotlike approach to questioning and to probing. Considerable danger can spring from the interviewer's parrotlike use of bookish questions or clumsy attempts to standardize questions. Since the same word has different meanings for different people and even for the same people at different times, the approach, the word, the statement, and the question must focus on the here-and-now needs and expressions of the client.

Often efforts at objectivity achieve a rigidity and a one-way action/reaction that ignores signs of anxiety or other cues to the emotional distress of the interviewer or the client. To disregard the client's feelings because the interviewer is determined to plod through a list of questions disconnects the interview situation from the client's problems. It is difficult to imagine how an interviewer intent on asking a list of predetermined questions is able to listen, show interest, and care for the client.

Freud was so concerned about the possible influence of the interviewer on the client that he encouraged the client to talk about anything that came into his mind (free association). The client reclined on the couch and the interviewer sat out of the client's view, behind the client. Direct questioning was sparse so as not to interfere with the client's flow of thought.

In the face-to-face experiential transaction, both verbal and nonverbal cues become even more important since there is more obvious sharing and experiencing of both thoughts/reasoning (cognitive aspect) and feeling (affective aspect). To note only outward behavior as expressed in words or actions or to structure interviews similarly for all clients neglects a most significant part of the process—the individuality of each client.

The effects of oversubjectivity may be just as deadening. The purpose of the interview is certainly not for both interviewer and client to share the same, or even different, crying towels. This does not infer that the interviewer must always conceal feelings but that the interviewer must not step out of his role and become the client.

The interviewer-turned-client may say:

- "I find so much pleasure in helping. I want so much to help you."
- "I'm lost, don't know what to do, am having such a hard time in this agency."
- "I have to have this interview with you or my supervisor will flunk me."
- "I know just how you feel. My father used to . . . ."

Oversubjectivity by the interviewer may incite the client to more frequent outbursts of emotion. These outbursts may be directed at the interviewer because of the tension built up by the interviewer's emotional distress. Sometimes it may be wiser for the interviewer to discontinue an interview if the emotions are too much to handle. Awareness of feeling is sensible. Expression of feeling (by the interviewer) must be carefully apportioned. Examples 6-5 and 6-6 describe some of the problems that may occur. Which of these examples exhibits the outside/inside difficulty?

Example 6-5
FIRST INTERVIEW

Today is the day! Carla's first interview! All of the class discussion seems a mad jumble at the moment. Carla has written down some opening questions as well as other kinds of questions that may help in keeping the interview going. She even remembered to include some ending phrases.

As she waits, she looks around the small office that has been given to her for this interview. She notes the littered desk and remembers how similar it is to the desk of her instructor. This gives her some comfort. Then the door opens and a tall, black man walks in.

CARLA: Hello. Please sit down over here. I want to thank you for volunteering to speak with me. I'm new here and am studying to be a mental health technician. *(Carla wonders whether she has said too much.)*

MR. B.: Yeah, Mr. S. told me. What do you want to know?

CARLA: Just want to know about you . . . about how you came to be in this hospital. *("Why did I say that?" thinks Carla. "All the opening questions I jotted down. No use.")*

MR. B.: Came to the hospital about a week ago, I guess. Really didn't want to come. Went out to Ed's house and we put a load on. Got home stoned; the wife started at me, yelling what a bastard I am, "Drunken bastard," wouldn't stop. So I began breaking things, maybe smashed her up, too. She called the cops; here I am.

CARLA: Yes? *(What do I say or do now?" wonders Carla.)*

MR. B.: That's it. Getting dried out so I can get out of here. Need to get back to work.

CARLA: Mr. B., please tell me more. I have to have more information about you. This isn't enough to make a report. *("Wow, that is a dumb thing to say.")*

MR. B.: Ask me some questions.

CARLA: Mr. B., you are being very kind and patient with me. Hmmmm. Tell me, when did you start drinking, what age?

MR. B.: Fourteen.

CARLA: Tell me more about how you began

to drink. *(Carla is frantically looking through her notes to find the question that should follow. She does not see Mr. B. clench his hand into a fist and then extend the fingers. Nor does she see the shifting of his legs from one position to another.)*

MR. B.: Drinking—everybody was drinking. At fourteen ran around partying and boozing. Later drank more and more to feel the high.

CARLA: I see. Did you have a happy home life? *(Carla thinks this has to be a good question. This question should always be asked according to a book she had read.)*

MR. B.: *(Looks at Carla with raised eyebrows. He bends down from the waist and cups his chin on his hands continuing to look at her.)*

Example 6-6
DIAGNOSTIC TURMOIL

Ms. M. walks into the office rapidly. She goes over to one of the chairs opposite Mr. P. and sits down, also rapidly.

MS. M.: I'm Missus M. I came to talk about my Henry. You asked me to come to discuss the tests you gave him.

MR. P.: Thank you for coming, Ms. M. I administered several tests to Henry to help me better understand how he is functioning. His teacher is concerned that Henry might have to remain in the first grade again.

MS. M.: *(Nods her head as Mr. P. speaks. When Mr. P. says that Henry's teacher had said Henry might remain in the first grade again, Ms. M. suddenly straightens herself so that her back almost seems rigid.)* Oh, no! Not stay back again. Take him out of school. Henry's a bright boy.

MR. P.: I guess I jumped too fast with that explanation. You're right, Henry is a bright boy. That's one of the things I found out.

MS. M.: *(Slumps a little bit in her chair and looks at her hands.)* What the? I—I—my husband is so angry, blames me, says I'm too easy with Henry.

MR. P.: Yes, I see how you feel. Concerned about Henry and about your husband's remarks. But what I have to say is really good news, and there is a way we can help Henry.

MS. M.: Good news? Then why does Henry need help?

MR. P.: From all the tests I gave Henry and from my conversation with him, I found out three things. Henry *is* above average in his intellectual ability. Yet he is upset because, although he is trying hard, he can't read. And, Henry needs some special help because he doesn't see things in the same way other boys his age see things.

MS. M.: Are you saying Henry needs glasses?

MR. P.: Glasses? Not as far as I know. What he needs is some special training to help him read more freely and comfortably. Henry shows what is called a perceptual difficulty. Have you ever noticed that he writes his b's and his d's backwards? Or, perhaps, have you seen him mirror-write? You know, it seems as if he is holding a mirror up and writing from right to left— letters backwards.

MS. M.: Maybe I did see that. I remember hearing a program on television once. They called this dys . . . dys . . . something or other.

MR. P.: Dyslexia. But labels don't tell us anything about Henry. The important thing is that we know how to help him now. And, besides, we found out about it early enough before damage is done to his education and to the way he feels about himself—and about learning.

MS. M.: *(Begins to sob softly and tears splatter down her face.)* Oh, what have I done? Tom was right; I was too easy on Henry.

MR. P.: Not you, Mrs. M. You did not do anything to cause this perceptual difficulty. However, you and your husband might aggravate the situation if you cannot accept Henry as he is and even help him at home with certain exercises. It's really not as horrible as you seem to think. I have dyslexia and am now studying for my doctor's degree. When I was attending school there was very little knowledge about perceptual difficulties. Today we have a good idea of how to handle it.

Missed/messed wavelength

The effort to find the client's wavelength of communication demands that the interviewer must first get his own self in order. Then the interviewer tunes in to the client's world. Lewis Carroll called attention to the need to hear the intended message rather than the words spoken by delightfully and insightfully pointing up the importance of searching beyond the words.

"But glory doesn't mean 'a nice knock-down argument'," Alice objected.

"When *I* use a word," Humpty Dumpty said, in rather a scornful tone, "it means just what I choose it to mean—neither more nor less."

"The question is," said Alice, "whether you *can* make words mean so many different things."

"The question is," said Humpty Dumpty, "which is to be the master—that's all."

Words do signify many different things. In order to be effective the interviewer must master communication so that both verbal and nonverbal messages are as clear as possible to him as well as to the client.

Background and experience barriers. Barriers to communication that arise from differences in background and experience are not always recognized by the interviewer. Even when interviewers and clients are from the same racial, religious, and socioeconomic groups, subtle differences in looking at the word may interfere with establishing a free exchange of ideas and feelings. Even when the interviewers have been able to recognize and largely free themselves from satisfying certain needs of their own, defending themselves in relationships, and expressing problems of their own life dramas—even then, unique differences are present in every interview.

The interviewer who has had experiences with many different people, has read, and has had encounters with various problems is one step ahead in developing an effective relationship. If the interviewer is able to honestly come to grips with the consequences of his own background and experience, several changes are likely to happen.

• Interviewers will learn to listen less selectively. The tendency to hear what one wants to hear because of the way one thinks will be avoided. Instead, the interviewer will listen to and observe the meaning to the client of the interviewer's as well as the client's words and behavior.

• Interviewers will expand their experience and increase their response repertoire so that they will have a larger store of communication content. Interviewers will be able to demonstrate more flexibility and better understanding in their communication.

- Interviewers will realize that both the interviewer's and the client's backgrounds and experiences are sources of discriminations. These discriminations will not only determine the level of communication effectiveness but also the manner in which new experiences will be accepted by the client.
- Interviewers will avoid the dim-sightedness of prejudices that tune them out of the client's wavelength. It is not easy to eliminate prejudices, but with practice, interviewers can screen them out for the duration of the interview.
- Interviewers will recognize that warped communication often results from words and behavior that are beyond the client's background and experiences. On the other hand, when the client's world is unfamiliar, the interviewer will not make a phony attempt to speak the lingo of the client. Instead, the interviewer will converse with (not to, about, at, or above) the client. The client is the only one who can convey whether the interviewer has tuned in to the client's wavelength.

The implication of these changes is that the responsibility for tuning in lies with the interviewer. Exercise 6-1 focuses on increasing awareness of the meaning of communication between two persons.

Exercise 6-1
TUNING IN TO IDEAS
AND FEELINGS

The group is divided into smaller groups of five, three people acting as observers and two as participants. One observer records nonverbal cues, one observer records verbal cues, and one observer records the entire process related to the people involved in speaking together.

The two participants discuss and tape a topic on which they have differing views. Each person says whatever he wants. Before the listener may voice his views, both the ideas and feelings of the speaker must be repeated by the listener and accepted by the speaker.

If the listener describes the speaker's meaning correctly, it is assumed that the listener heard and understood the speaker. If the listener's description is unacceptable to the speaker, either the listener placed obstacles in the way of understanding or the

speaker did not make himself clear. By this process of listening, repeating, and speaking, both the listener and the speaker are given opportunities to clarify their thinking, to examine their prejudices (obstacles to understanding), and to concentrate on what is being said rather than on the reply to be made.

After the tape of the discussion is played back, all five members of the group discuss the level and adequacy of communication.

Perceptual blur. Background roots and experiences affect the way individuals treat differences. Sometimes the pretense of non-bias works in the guise of tolerance. However, in the close emotional relationship of an interview the deception is frequently revealed.

People may think that to be tolerant is great. It makes them feel so good and liberal to tolerate that "other-race" family who just moved in on their block, that radical who is disenchanted with the political scene, or that person whose religious ways are peculiar.

Tolerating differences is not the same as accepting them. Accepting differences implies nonconditional respect. This differentiation between tolerance and acceptance is not just word-maneuvering. The intent of these sentences is to bring awareness to the contrast between *tolerance*, which is judgmental, and *acceptance*, which is nonjudgmental.

Tolerance suggests that one individual indulges or permits another individual to maintain certain beliefs or behavior. Tolerant people ("we") see others ("they") as fleeting perceptual blurs. They perceive neither themselves nor others clearly or honestly. A tolerant person endures the peculiarities and differences of others.

In contrast to tolerance, acceptance embraces ideas and feelings that support the worthiness of each individual with no judgmental gradations of better or worse. This does not suggest that acceptance is passive or neutral. Acceptance does not necessarily express agreement or approval. Acceptance is a complex of attitudes that consist of a number of characteristics (Table 3).

Before beginning an interview, the characteristics of acceptance should be clearly understood. These characteristics are expressed more in nonverbal than in verbal

Table 3. Acceptance: what it is and is not

WHAT IT IS	WHAT IT IS NOT
Respect (positive regard and caring) Client's self-experience is *real* to him. No experience should be considered more or less worthy of positive regard. Interviewer maintains attitude of warm goodwill (calmness and understanding) regardless of what client talks about, even if it is not socially acceptable or to the interviewer's personal liking. (Respect or liking is expressed in spite of client's unlikable characteristics.) Client is valued because of his aliveness ("He is") and his being ("He thinks and feels"). Client is a person with dignity, not an object to be pulled apart for study. Interviewer believes that there is validity in the distinctive values of dissimilar ethnic, religious, and socioeconomic backgrounds. (Each person has the right to feel, think, and value differently.)	Agreement Approval Neutrality Cold detachment Idle and/or avid curiosity Thinking or feeling the way client does or having same values Assessment of client's values
Client-directed Client is encouraged to make his own choices and to determine his own life, providing he does not infringe on the lives and rights of others. Interviewer encourages client to be his own source of self-evaluation while affirming client's worth and potential to become a fully functioning individual who is open to experience.	Changing client in interviewer's image Disregard of limitations of external reality Client submission to controlling interviewer
Noncritical kindness Because of assurance of complete confidentiality, client feels safe in discussing any topic he chooses. He feels free to relate experiences that he may be ashamed of or that frighten him as well as those of which he is proud and satisfied. Client feels accepted by interviewer while still feeling unacceptable to himself. Interviewer expresses continuing willingness to help no matter what the behavior of the client and no matter whether he approves or disapproves of client's behavior.	Avoidance of discussion of alternative behaviors for client's consideration Judgmental Blaming person Releasing client from responsibility for his acts Allotting all responsibility to client for his acts Condemnation/punishment Hostility Rejection
Genuine interest Interviewer is concerned about meaning of each item of discussion to client's welfare and comfort. Interviewer constantly tries to clarify his own understanding of client and to communicate that understanding during the interview. Interviewer recognizes and takes pleasure in client's achievements, expresses confidence in client's ability to handle certain tasks or situations, and honestly appreciates client's efforts.	Evasiveness Distorted interpretations Disregard of appropriateness of client's way of functioning Unrealistic appraisal of client's capacity Insensitivity to client's conception of his abilities
Empathy Client feels trust in interviewer and therefore is able to ventilate (express) his feelings. Interviewer is able to participate in client's expression of feelings; he assures client of naturalness of anger as well as of joy. Interviewer is able to see events through eyes of client.	Feeling uncomfortable with client's outburst Fear of not being able to control client's emotional expressions Oversolicitude Sympathy (too much emotional involvement by interviewer)

Table 3. Acceptance: what it is and is not—cont'd

WHAT IT IS	WHAT IT IS NOT
Human equality (all people create egos that are equally worthy)	
Client is a co-worker on a common problem. Client's opinions and feelings are worthy of consideration.	Interviewer's advice as primary direction for change
Interviewer conveys feeling of quiet friendliness as well as of strong desire to help client.	Extreme cordiality and effusiveness
Effective communication through understanding	
Interviewer's comments are on the same wavelength of that of client. Interviewer follows client's line of thought.	Interviewer attends to his own line of thought because he knows the score
Interviewer attempts to anticipate events the way in which client anticipates them. Interviewer communicates this understanding according to client's way of thinking while still maintaining a professional overview of client's problem.	Interviewer anticipates events according to his theoretical outlook
	Tries to fit client's behavior into a statistical pattern

Table 4. Acceptance: how it is shown by interviewer

NONVERBAL BEHAVIOR	VERBAL BEHAVIOR
Facial expression	*Verbal following behavior**
Relaxed, not frowning	Appropriate choice of words and statements that follow client's comments as well as nonverbal cues
Eye contact	
Visual interaction not forced	Avoids excessive interviewer talk time
Looks at client but does not stare intensely and makes varied use of eye contact	Feedback to client of an undistorted communication of what client said
Body posture	*Free client verbal expression encouraged*
Comfortable posture, relaxed movements, and "towardness" posture conveyed by slightly leaning forward or sitting forward on chair	No interruptions; reinforces client's verbal expressions by nonverbal signs such as smiles, nods, and body posture
Gestures	*Language used according to client's level of understanding*
Loose and natural arm, hand, leg, and foot movements	Words used to aid understanding not to impress; use of special words (colloquialisms) of client only if they are understood by interviewer
Distance	
Comfortable distance face to face	*Congruence between verbal and nonverbal behavior*
	What is said fits nonverbal expressions; interviewer looks like he means what he says

*Ivey (1971) describes verbal following behavior as one of the behavioral characteristics of attention, which is a crucial reinforcer encouraging the client to continue talking.

behavior. Table 4 explains acceptance in behavioral terms.

Misdirected motivation. Beyond the characteristics of background, perception, and acceptance, the effective interviewer must be on the lookout for traps set by misdirected motivation.

Defensive motivation. In interviews the "knowing" person offers his understanding while seeking to assist the client find satisfaction and resolutions to his problems. Yet one immense barrier to such transactions might arise from defensiveness. One or both participants in the interview may set up a barrier to hide an undesired revelation. If the interviewer is defensive and concerned about concealing his self-centered motives, then the trouble starts.

In order to appear intelligent, virtuous, capable, and correct, the interviewer may manipulate the conversation to revolve about certain safe themes. The outcome may be a sorting of messages. This selection, rejection, and evaluation results in a more stilted, less acceptant, and less honest atmosphere. The interviewer is more likely to be overrestrained in his messages. The interview becomes heavily ladden with distorted motivation misdirected to the interviewer rather than the client's needs.

Buddy motivation. Sometimes, in order to induce client motivation, the interviewer falls back on buddy motivation. The interviewer seeks responses from the client by offering friendship through an exchange of problems. At times this may amount to a barter: "Here's a problem (or confidential matter) for you. Now you tell me something."

Novice interviewers are especially prone to depend on a misconstrued conception of establishing rapport (harmonious relationship). The new interviewer frequently believes rapport is another word for friendship or for being a buddy. One jarring problem emerging from the establishment of a friendship is that such a status may encourage clients to withhold their true feelings in order to avoid hurting their new-found friend, the interviewer. Clients may distort their own reactions to conform more nearly to what they think will please the interviewer.

Friendship and friendliness have different implications. *Friendship* suggests joining with another in a kinship of mutual intimacy and a free exchange of values and understandings, of thoughts and feelings. *Friendliness* indicates certain characteristics suitable for friendship, such as intimacy, free exchange, support, help, and dependability. The primary difference between friendship and friendliness resides in the degree of intimacy and free exchange. Friendship veers off into a "no holds barred" situation in which each individual opens up without any interviewer "stops" of intimacy and revelation. Such a relationship is usually pleasing rather than releasing. In contrast, the friendly interviewer has an awareness of the client's needs and adjusts his intimacy and friendly exchange to the client's needs.

Nagging motivation. Another kind of inap-propriate motivation develops when the interviewer tries too hard to establish motivation. Too many questions are asked; the client is asked over and over again what is bothering him, why he won't talk, and what is wrong with the interview. On and on the interviewer presses to get more responses and lengthier explanations from the client. When the client senses this nagging pressure, he may indicate an impatience by requesting that the interviewer stop pushing and questioning so much, being so nosey, and get on with it. Another tactic the client may use is to clam up. A stalemate may result in which whatever motivation the client may have originally had for seeking help may be decreased. Even greater undermotivation may appear. When the client becomes silent and reduces or even discontinues *working* on his problems, the interviewer had best examine whether the interview is indeed meaningful and rewarding to the client.

●   ●   ●

Unless the client wants it to, very little will happen. The nature of the interviewer's motivation is one aspect. The encouragement and building of the client's motivation is the other aspect. Misdirected motivation in the form of defensive motivation, buddy motivation, or nagging motivation may satisfy the interviewer's needs, but these rarely encourage client participation in an honest partnership of self-discovery and self-growth toward behavioral change.

The interviewer's work is to be an enabler not an enforcer of change. He assists in removing the blinders that keep the client from facing his predicament honestly. Although the interviewer extends a helping hand, the client controls whether he will accept the hand. Of particular importance is the client's discomfort with his present plight (degree of cognitive dissonance). The more the client is reinforced by himself and/or others in his crutchlike behavior, the less he feels the need for change.

Loss of cues. Cues are the subtle sources to which both the interviewer and the client react. These may be such observable characteristics as skin color or facial scars or the more subtle individual gestures or vocal features to emphasize or deemphasize what is said.

From these cues evolve expectations regarding the client's apparent economic status, probable attitudes on certain topics, and level of information in certain fields. Unfortunately these cues may be distorted out of proportion while other cues are ignored. How these cues are interpreted depends solely on the viewer. As a result of an individual's selective perception and interpretation of cues, insurmountable restrictions may clog the flow of communication and understanding.

Exercises 6-2 to 6-4 are aimed at sharpening one's awareness and decreasing the degree of selectivity of perception.

Exercise 6-2
THE CUES AROUND EVERYONE

This exercise may be done individually or in small groups. Each individual shuts his eyes and listens for sounds for 60 seconds. Each individual then opens his eyes and jots down the sounds he remembers. After 1 minute of writing, the participants in the group read their list of sounds. Group members compare individual lists in terms of the total number of sounds heard and the kinds of sounds heard.

After 5 minutes of discussion and comparison, each individual shuts his eyes again and listens for sounds. During the group discussion that follows the second period the number of sounds, kinds of sounds, and changes in number and kinds are noted.

Exercise 6-3
HEARING "FAST" AND "SLOW"
"Fast" hearing

Five people are involved in this exercise. Four people are observers; the fifth person talks into the microphone of a tape recorder as rapidly as possible. The speaker may discuss any topic, including himself, for 5 minutes. The four observers take notes about nonverbal cues, verbal cues, the "whole" person talking, and what the speaker is saying, one observer for each kind of information.

All observations are recorded. A 5-minute group discussion follows in which the observers and the speaker compare notes about their recorded observations and the speaker's feelings about the entire process. Discussion should include what the observers think they may have left out. Then the tape recorder is turned on and notes are taken by all five members of the group of the cues on the recording. An additional 5-minute group discussion is conducted concentrating on the added cues derived from the tape.

"Slow" hearing

The entire procedure for the "fast" hearing exercise is repeated, the only difference being that this time the speaker discusses a topic as slowly as possible.

Exercise 6-4
"SPEAKING" WITH NONVERBAL CUES

The group pairs off with partners facing each other. For 1 minute both partners try to tell one another something without speaking. They then discuss for 5 minutes what each was trying to communicate. The procedure of nonverbal "speaking" is repeated for another minute. Once again the partners discuss what they were trying to communicate. It should be noted whether either or both partners improved their ability to communicate. If so, one should explain what caused the improvement.

There is no doubt of the importance of developing accurate cue perception. This importance extends beyond the interview transaction into the daily lives of people and farther into the international scene. The loss or distortion of cues too often results in a feedback loop that increases to the point of monster proportions. Each new misperception is added to the previous one until all perspective is befuddled.

A vicious cycle ensues that cannot be broken until the individual involved in this cycle becomes more skilled in picking up cues and less prone to permit distortion. Fig. 6-1 diagrams this feedback loop of cue distortion.

Another danger of cue loss or distortion can come from errors in recording. Several possibilities can crop up that push the interviewer into a twisted round of errors. Perhaps the most frequent misjudgment appears when the interviewer hears only what he expects the client to say. These warped statements are often rounded out, amplified, or otherwise modified by the interviewer to fit his preconceived notions. Furthermore, the interviewer may embellish these im-

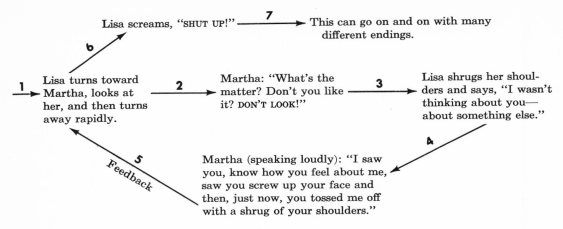

Fig. 6-1. Feedback loop of cue distortion.

perfect recordings with a change in the sequence of what the client says in order to make the continuity more logical—to the interviewer. This may be a very serious infraction since the client's hopping from one subject to another may be a distinct characteristic of his behavior. Other forms of incorrect recording may be the omission of selected statements of the client that threaten or run counter to the interviewer's own attitudes or to the interviewer's belief in the progress the client should be making.

Recordings that are finished pieces of composition are unlike actual conversation. Conversation does not usually flow in a straight line from thought to thought (linear communication) but actually hops from idea to idea centered about a theme. Thus the interview is actually a hodgepodge of thoughts from which the interviewer must glean the themes revealing the client's concerns. Polished recordings tend to distort cues. These refined recordings are just the figments of the interviewer's pen.

Exercises 6-5 and 6-6 are to be used for the purpose of sharpening one's seeing and hearing.

Exercise 6-5
GARBLED LISTENING

The people in the group arrange themselves into two circles. The inner circle should contain half the number of people as the outer circle. Someone in the inner circle starts a one-sentence statement (rumor) about the outer-circle members and someone in the outer circle starts a similar one-sentence statement about the inner-circle group. Any individual may begin the rumor with either a positive or negative remark stated in one sentence.

The starter writes down the rumor on a piece of paper and then whispers it to the person on his left. This next person writes down what he heard and then whispers it to the next person. This procedure of writing (concealed) what is heard and then communicating it to the person on the left continues around the circle until the starter has been told the rumor and has written it down below his original statement.

When the two groups have finished there should be a discussion of the various changes in the original statement as the rumor moved from one person to another. The discussion should also include the degree of difference in distortion (changes from original statement) by the smaller inner circle as compared to the larger outer circle.

After the discussion, which should take no longer than 10 minutes, the same procedure should be repeated with a new starter and a new rumor.

Compare the recorded whispers in terms of their accuracy to the original statement from the first to the second whispering campaign.

Exercise 6-6
BUNGLED VIEWING

The group is divided into pairs. The chairs of the partners should be arranged so that

	FIRST OBSERVATION	SECOND OBSERVATION
Number of accurate cues		
Number of omitted cues		
Number of inaccurate cues		
Number of interpretations		

the two people are seated back to back. Each person has 5 minutes to write down what the other person is wearing and what the other person looks like in general. The observation should include a general description of the other person, with specific details about clothing. The description should be focused on what the describer actually recalls seeing, not how the describer feels about the person.

After both participants have completed their descriptions, they discuss their notations. The partners face one another now and note the degree of accuracy and completeness of their observations.

• How many cues have been recorded?
• How many cues have been omitted?
• How many cues are inaccurate?
• How many notations are interpretations or express feelings rather than observable facts?

The accuracies, the omissions, the inaccuracies, and the emotionally toned observations should be tabulated as suggested in the chart above.

The entire exercise is repeated with another person. Then a comparison is made of the extent of improvement from the first to the second observation, as indicated in the two columns of the chart.

Low level of communication skill. At the beginning of this discussion of missed/ messed wavelength, the "great philosopher" (?) Humpty Dumpty was quoted as saying that words meant just what he chose them to mean. Mr. Dumpty sacrifices communication to his own glorified choice of meaning. He may be the master of the word but is not equal to the task of conveying his meanings. Yet Mr. Dumpty is a symbol representing the kind of communication that is prevalent in the discussions of most people. He reflects the self-centered (narcissistic) communica-

tion that occurs too often in interpersonal discussions. All too often speech conceals rather than communicates.

Since it is through communication that people relate to one another, the ability to adequately communicate is particularly important for the interviewer. By both verbal expressions and nonverbal gestures plus grunts, ahs, ughs, hmmmms, and pauses, the interviewer sends messages to the client and receives return messages from the client. How clearly the interviewer sends and receives messages depends on his appreciation of his own and the client's background and experiences. The differences that do exist need to be recognized rather than swept under the cover of "I didn't even notice that. . . ." Connected to this recognition must be acceptance of the client's *right* to be different, to be himself.

As the interviewer begins to really see and honestly accept the client, communication becomes easier since both the client and the interviewer value themselves as important to what happens at any one interview moment. This feeling that each person is worthy motivates each one to speak *with*, not *to* or *at*, the other person. For the interviewer, this means developing the skills of the "what" and the "how" of communication.

The "what" of communication obliges the interviewer to recognize where the client is intellectually, verbally, and philosophically. Simple messages strengthen communication. The vigor of this simplicity comes from its directness and its freedom from double meanings.

Simplicity is not that simple, however. What's simple for one person may be overly simple for another. Example 6-7 shows two contrasting views of communication that point up different approaches to getting a message across.

Example 6-7
TRANSLATING THE COMPLICATED MESSAGE

Complicated message	Translation
Her esophageal contractions emitted decelerating auditory responses as a 5-minute schedule of negative reinforcement was presented.	She began to slowly stop talking when Tom called out "shut up" every 5 minutes.
Smart Sam pools the bread and runs out to score some barbs. He brings back and shot up five Mexican Reds. Sam got offed.	Smart Sam collects the money and runs out to purchase some barbiturates. He brings back and injects a mixture of 20 mg secobarbital and 10 mg strychnine. Sam dies.

Effective communication demands an atmosphere of trust and respect in which the meaning of both verbal and nonverbal symbols are familiar to the participants. The interviewer is responsible for the kind of communication that is established. He must choose words that express his meaning to the client and that are in tune with the economic class, education, ethnic background, age, geographical region, and other factors of importance to a particular client.

There is one important caution. When the interviewer adjusts *what* he says in accordance with the characteristics of his client, he must avoid the phoniness of trying to talk in the client's vernacular if this way of speaking is not comfortable for him. The interviewer must find the middle ground in which certain words familiar to the client may be interjected to show that the client's language is accepted and understood. What is even more important than speaking in the client's words is that the interviewer understands the meaning of the client's facial expressions, tone of voice, posture, and gestures. While the interviewer is accomplishing the skill of *what* to say, he must also look to *how* to send these messages.

The "how" of communication concentrates on the interviewer's skill to:

1. Maintain a minimum level of interruptions of the client's flow of conversation
2. Hear what the client says and not speak the client's answers
3. Avoid loaded questions that maneuver the client into the interviewer's thought patterns
4. Refrain from negative statements such as "I don't suppose you thought about . . .?" or "You didn't . . ." or "Sorry about that"
5. Pace the speed and rhythm of his communication so that it is not too slow, too fast, too much, or too little

In addition to these five points about verbal content, the interviewer must be alert to his own nonverbal cues. Preoccupied with his own problems, the interviewer may discourage the client's spontaneous discussion. The body-shifting, clockwatching, sighing interviewer is playing his own game in which the client is an intruder. Such an interviewer must solve his own problems first.

Opening the interview
The initial greeting

The where and how of greeting a client may come easily to some beginning interviewers. Others may quake at the thought of the client's first visit. The interviewer may wonder where the client should be greeted—in the reception room, as the client enters the interviewer's office, or after the client has been seated?

How should the client be greeted—with a handshake, with an introduction to the interviewer, with small talk, with "hello" or "hi" and the client's name, or with a nod and/or a smile? Should the interviewer be seated or standing when he greets the client? Should the interviewer walk toward the client? What name should the interviewer use for the client? Should the interviewer greet the client by his first name or by his last name? What name should the interviewer use? Should the interviewer introduce himself or herself as Mr., Miss, Mrs., Ms., first and last name? Should he use his title and his name or should he use only his first name?

The answer to all these questions depends on the way in which the interview is initiated, the life-style of both the client and the interviewer, and the age of the client.

Initiating. A hearty greeting to a depressed client and a strong handclasp for a 6-year-old are obviously foolish. Such foolish greetings are possible when the interviewer is self-occupied and/or uptight. The interviewer's immersion in his own problems prevents him from heeding the client's cues for a greeting.

A dominant point influencing the nature of the greeting arises from how the interview was initiated. The approach to the client-initiated interview is somewhat different than the other-initiated interview, particularly if the client is compelled by the courts, a school, or a parent to meet with the interviewer.

When the interview is client-initiated, the interviewer is less likely to have to deal with lack of interest, piecemeal cooperation, or outright hostility. The task of beginning is simplified if the client initiates the interview. Then, either the client begins to discuss the problem that brings him to the interviewer or a simple suggestion from the interviewer will spur him on, "Would you tell me what brings you here?"

If the interview was arranged by someone other than the client, the interviewer tries to discover who the client thinks made the request and what the client believes urged this person to refer the client. Since the interview is not client-motivated, the interviewer will have to work harder to begin the session. Some possible beginnings for this nonclient-arranged interview are given in the following examples.

1

To the 17-year-old whose parents have asked about some educational directions for their son: "Hello, I'm Dr. Y. I am going to help you make some decisions about your educational plans."

2

To the adult who has been asked to speak with a social worker about where she would live after she leaves the hospital: "Mr. H. was talking to me about your concerns. Let's explore the choices you have."

3

To the 16-year-old girl who has been arrested for shoplifting and uncontrollable behavior: "I've been asked to work with you. I know it's tough to be reporting to me regularly. Together we can work something out which will be helpful. Let's try."

4

To the angry client who has been brought to the interviewer against his wishes: "I can understand your anger since you did not want to come here" or "I gather you feel you were forced to come here against your wishes."

No one can borrow another's words. The suggestions for beginning are just that—suggestions. Each person must decide on his own words.

Exercise 6-7 offers an opportunity for experiencing some of the methods for beginning.

Exercise 6-7
INITIAL GREETINGS

The class is divided into groups of four. Each individual tapes beginnings to one of the situations cited below. The same person plays the role of both the client and the interviewer and tapes both of these roles. During the taping three other trainees observe and record the client-interviewer role-player's verbal and nonverbal behavior. The beginning greeting should take no more than 2 or 3 minutes for each situation.

A person selects a situation and assumes for 2 minutes the roles of both client and interviewer. His role-playing is taped. During his role-playing he is observed by the other members of the group. Observations are recorded. After this first role-playing the tape is played and a 5-minute discussion is held on the positive aspects of the beginning. The role-player should include how he felt in the dual roles in this discussion. Then all members of the group make suggestions for improvement.

The same person repeats a 2-minute role-playing of client and interviewer, which is observed and recorded by the other three group members. Once again the interview is taped, and a 5-minute discussion follows. This second discussion should begin with improvements that were made in the greeting and how the role-player felt in the dual role. Suggestions for improvement are then made. This procedure is repeated for the other three members.

Situations for role-playing

1. Tom walks into the personnel office of the large department store holding a copy of the *Daily Post* in his hand.
2. The woman is led into the quarterway house for alcoholics. She walks slowly and clutches a handkerchief to her lips.
3. Tisa, 4 years old, is pulled into the

doctor's office by her mother. Tisa's face is wet with tears.

4. Manny pushes past the secretary, who is announcing him to the principal, and says, "Okay, so here I am. Whaddya want?"

Life-style. Another challenge in the early moments of the interview is the client's life-style. An individual's life-style originates from the way in which his parents viewed the world and develops in accordance with the individual's unique experiences. Life-style includes the values, attitudes, and behavior that an individual learns.

The conduct of a successful interview from beginning to closing must be in balance with the life-style of the client as well as with the life-style of the interviewer. Some individuals are comfortable with small talk about the weather or about an item in the interviewer's office as "ice-breakers." Others might view this small talk as disrespectful. Small talk may encourage some clients to escape their immediate problems with more small talk. As a result, the ice-breaker becomes a barrier to communication.

The client's life-style also will affect his expectations of the interviewer's office, the interviewer's appearance, and the interviewer's approach. These expectations may be revealed by such comments as, "Where's your couch?" or "You don't look like a head shrinker." The interviewer who pursues the client's line of thought may open up some fruitful channels for further communication by replying, "No couch. Don't use one. How do you feel about rapping on these comfortable chairs?" or "What do you think a head shrinker looks like?" What the client says or does not say in the first few minutes is an effective jumping-off place for what will take place during the rest of the interview hour.

The client's life-style provides the compass to direct how and which details he presents and how and whether he will reveal the precipitating stress that stirs him to seek help. The interviewer's life-style influences the way he encourages the client to bring forth or hold back his thoughts and words.

The interviewer who inspires the client in this beginning stage to believe "You can do it," "There are solutions," "You are

capable" facilitates the client in inspecting his feelings and experimenting with new approaches to solving his problems.

No matter what the client's life-style may be it is the interviewer's responsibility to offer the client instant assurance of his genuineness and interest. Even the most willing, eager-to-do-something client feels some degree of fear, despair, or confusion about what might happen during the interview.

It is up to the interviewer to weigh all the cues from the client and then decide whether this particular client is ready for a handshake or would be disturbed by any form of body touching. It is up to the interviewer to treat the client with the same humanity as a guest welcomed in his home. Exercise 6-8 focuses on beginning interviews with people with different life-styles.

Exercise 6-8
RESPONDING TO LIFE-STYLES

The class is divided into groups of no more than five members. Each trainee writes two replies for each of the four situations presented in the chart on the opposite page. In the first column the trainees write replies that indicate a lack of awareness and lack of acceptance of the client's life-style on the part of the interviewer. In the second column the reply should show recognition and acceptance of the client's life-style. The replies should be discussed by the five group members; the preferred replies of the small groups can then be discussed by the entire class.

Client's age. The developmental changes in an individual as he ages are one cue for the initial approach to the client. The younger the client, the less the interviewer can depend on verbal cues. The body also "speaks" to the interviewer. Body posture and movements can provide important nonverbal cues. Not only children but also adults tell a great deal by means of these nonverbal cues.

With children, the initial gesture may have to be a nonverbal greeting such as offering a toy through which the child may "speak." Finger, hand, and string puppets are excellent vehicles for conversation for the interviewer as well as the child.

The younger the client, the more likely that help was sought *for* rather than *by* him.

STATEMENT ABOUT CLIENT	INTERVIEWER'S NONACCEPTANCE	INTERVIEWER'S ACCEPTANCE
May, 14 years old, comes into the office carrying the book *Don't Shoot, We Are Your Children* (Lucas, 1972). May has been in the detention home for girls for the past 2 weeks. She begins to speak as she enters the room, "Mixed-up place. All I know is don't want to stay here anymore. Not sure about anything."		
Stan sits down in the comfortable chair upholstered in orange imitation leather. He looks around the office and notes how sparsely it is furnished. His hands are trembling and his body twitches as he moves around. "Listen, Doc, cut out talk-crap. Drink a lot. Gotta stop. Ain't nuthin more to it."		
Edna, a 19-year-old black woman, was admitted 3 months ago for attempted suicide. A quick smile spreads over her face as she says, "Mr. T., last week was hellish. Looked around for a razor blade again. My mother brought my baby. Mom cried. Has no money for herself. No money for the baby's milk. Too much."		
Jeff bursts into the office. He carries a large sign that reads "Rally for Angela. Down with the pigs."		

Therefore the two questions of interest to the interviewer—"Who felt the need for help?" and "Why now?"—must be answered by someone other than the client.

Setting limits

At the beginning of an interview there also must be some mutually approved ideas about setting limits. The helping road is much smoother if both client and interviewer agree on a cooperative pact that is both satisfying and realistic for the interviewer as well as the client.

Time limits. The security and ease of both client and interviewer are enhanced by an informational exhange about the time of appointments, what happens if appointments are to be missed, and how tardiness is to be handled. Time requirements that are too loose or too rigid disturb the effectiveness of the interview relationship. There is no fast-help, one-shot approach to finding a middle ground between too little or too much. The interviewer must adjust the time specifications in accordance with his own and the client's needs as well as with the client's cultural framework.

Time is judged and handled differently in various cultures. For instance, the pressure of time and getting places *on time* means more in the Western than in the Eastern cultures. Time means more to the harried businessman who wants to increase production and less to the youth of the counterculture who is turned off by the materialistic surgings of his elders.

The conditions of life make for different perspectives about time. Delaying gratifications becomes a burdensome chore for the black person, the Indian, and the Chicano who are grappling with the present satisfaction of their needs. Telling people of these cultures "You will feel better in time" often discourages their faith in the interviewer.

For people from these cultures, longer periods for each interview with some apparent positive results are far more effective than interview sessions conducted over a longer period of weeks. Time has different meanings for the prisoner who is serving an indefinite sentence than the one who has knowledge of the length of time he has to serve, for the anxious person who tends to overestimate the passage of time, and for the older person who is doubtful about the future and tenaciously holds on to the present.

The interviewer who stands aloof from the world of his client loses out in setting time limits. Unfortunately all too often interviewers have been trained to function with people similar to themselves and are not prepared "to deal with individuals who come from different racial, ethnic, and socioeconomic groups whose values, attitudes and general life styles may be at great variance" with their own (Wagner and Haug, 1972). These authers also concluded:

A minority group individual might find himself in the difficult position of trying to overcome what Schofield (1964) describes as the therapist's "Yavis syndrome,*" a common tendency of the therapist to differentially select clients who are successful, young, attractive, intelligent, well educated, verbal, and introspective."

What this discussion is leading up to is the frequently repeated phrase "Time is relative." Therefore, although time limits are a necessity, considerations of age, ethnic background, and immediacy of problems as well as the impact of the client's present situation are the foundation for arranging time provisions. Time limits exist not only to fulfill the interviewer's schedule but also for the client's advantage.

The client's attention may have to be drawn to a briefer period of time on a certain day; for instance, "Ned, we have 30 minutes today; let's see what we can accomplish." Toward the end of an interview session, the interviewer may say, "There are 15 minutes left; let's try to pull together some of the things we've talked about."

Example 6-8 suggests some responses when the client is late.

*The term "Yavis" comes from the first letters of young, attractive, verbal, introspective, and successful.

Example 6-8
WHEN THE CLIENT OR THE INTERVIEWER IS LATE

1

The first time the client is late the interviewer listens to client's explanation without comment. Approving comments such as "That's okay" may reinforce the client's behavior. In fact, the client may begin to wonder whether 45 or 50 minutes are really necessary for the interview or perhaps whether the client is really important to the interviewer.

2

After the client has been late once, the interviewer may say, "Sorry to hear about your clock conking out again. We will do as much as we can in the remaining 35 minutes."

3

When the client says, "Forgot about my appointment until it was time to leave," the interviewer's reply might be, "How did you feel when you realized you would be late?" or "I wonder, did you feel as if you really did not want to come?" If the client's answer is "yes," the interviewer helps the client explore his feelings. If the client's answer is "no," the interviewer holds off further discussion for another occasion. The client is probably not ready to talk about his feelings at the present moment. Responsibility for the length of the appointment and for promptness must be shifted to the client.

4

When the interviewer is late, he should make adjustments so that the client will not lose interview time. It is the interviewer's responsibility to either lengthen the interview session or make up the lost time at a later session, unless the client prefers not to do this.

The client's early arrival may reveal some degree of anxiety just as his late arrival may signify a reluctance to participate. If the early client shows by word or by gesture that he feels the interview is threatening, the interviewer reassures him by saying, "Glad to see you, Mr. A. Please wait in the reception room. I'll be with you at three.

The client in my office will be finished by then." The interviewer's smile and nod also provide needed support.

Sometimes the client plays with time by making frequent requests to change the time of his appointments. The interviewer's response depends on the client. For one client, changing the time occasionally may reassure the client of the interviewer's interest in him. For another client, providing such changes may encourage the client to attempt other manipulative devices. For still others, the interviewer would appear as disorganized as the client and thus lose face with the client.

When the client asks, "What time is it?" at frequent intervals, the interviewer should be aware of the possibility that the client may be bored, annoyed, withholding information until there is too little time to discuss it, comparing the present interviewer's procedure with a past one, or any number of other possibilities. Some productive interviewer comments are, "You seem to be concerned about the time. Is there something you want to do?" or "Is there something you need to tell me about?"

The words must be individualized. The interviewer must look at what is happening between the client and himself and use the preceding ideas as guidelines for his responses. Time limits as well as other limits should be minimal and in accordance with the client's belief system. In addition, these limits should be applied nonpunitively.

Behavior or acting-out limits. The interviewer may be puzzled about the when and how of setting limits on the client's behavior. Reluctance to assert himself or inability to establish a framework of acceptable behavior may impede the interviewer from establishing behavior limits. The important idea in setting acting-out limits is that it is acceptable for the client to *tell about* how he feels and what he would like to do, but there is certain behavior that *cannot be acted out.*

Some verbal comments may not fall easily on the dainty ears of some interviewers even though the spontaneous release of the feelings may help the client over a barrier he is experiencing. When the client shouts "Man, you're a son of a bitch! You made me remember . . .!" his explosive comment may open up a wealth of more relevant thoughts and feelings. The interviewer who becomes defensive about such tongue-lashing reveals this reaction by a comment such as, "Who do you think you're talking to? Cut out the foul language." On the other hand, the interviewer may compensate for his discomfort by interpreting the client's outcry in the following manner, "Aren't you feeling the same way about me as you did about your father?" Such an interpretation may shut off the client's expression of feeling because it is untimely or incorrect.

It is different, however, when the client approaches the interviewer in a menacing fashion or raises his hand as if to strike the interviewer. At such a moment the interviewer immediately must stop the client. Stopping the client from acting out his hostile behavior is not only for the interviewer's protection but also to rescue the client from his own guilt feelings. For instance, the interviewer says, "Mrs. T., I'll listen to anything you want to say—how you feel about your brother or about me. But I won't let you hurt me," or "Bobby, it's okay to tell me you hate me, but I won't let you break that car."

The adult as well as the child client most probably is hoping to be restrained from acting out his hostile impulses. Permitting the client to hurt or break something often arouses feelings of shame and increases rather than decreases anxiety feelings. The client is encouraged to express his feelings verbally. He is restrained from expressing his feelings through action.

Process limits. A third kind of setting limits is in part concerned with the client's expectations from the interviewer and in part with the degree of dependence the client desires and sometimes the interviewer encourages. Process limits refers to the **degree** to which the client and the interviewer are participating in the interview transaction.

The client may await the interviewer's questions or remarks and put forth little or no work effort in the interpersonal transaction. Another client may talk too much or ask continuous questions, seeking directions and answers from the interviewer. Symptom listing becomes the chief method of some other clients.

Example 6-9 provides samples of some problems and solutions for process limits.

Example 6-9
PROCESS PROBLEMS

THE OVERDEPENDER: "Wow, you really want me to tell you what to do. I bet together we could work out a good arrangement for your apartment searching."

"I know you want me to tell you how I would act in this same situation. But that would be *me, not you*. Do you really think that it would help you if I spelled out a step-by-step approach to the problem?"

THE CONCEALER: "Lee, I'm here to help you. Unless you come forth with your feelings, your ideas, how can I help?"

THE WAITER: "Cindy, tell me about the incident. Asking you questions is a drag. You're the only one who can tell it like it is to you."

THE SYMPTOM RECORDER: The client lists symptoms, reviewing them in great detail. The interviewer guides the client into a discussion of one of his problems. "Mr. T., you have told me a great deal about your daughter's horrible behavior and your stomach pains. Let's look a little more closely at your feelings about your daughter now."

THE MERRY- (OR SAD-) GO-ROUNDER: The client begins to talk as soon as he enters the office and rambles on about problems, all dominated by a central concern that he repeats like a broken record. The interviewer picks out the recurring theme and says, "Over and over again people seem to tell you to shut up and to mind your own affairs. How do you think you might change what's happening?"

THE ROLE REVERSER: The client asks the interviewer personal questions about his life. The interviewer answers if the question is pertinent to the client's problems and/or the interviewer's reply will indicate empathic understanding of the client. If the client continues the personal questions, he may be trying to avoid his own problems. In this instance the interviewer might say, "I could go on and on talking about myself. I do think we ought to get on with your distress about the people who are working with you at Xaviar Company."

THE PRAISER: The client persists in complimenting the interviewer and repeats such statements as "I know you're the only one who can help me" and "You must know the answer, you're so capable." The interviewer answers, "Your confidence does make me feel good. I do know, however, that without your help little can be accomplished."

THE SELF-CONTEMPTER: The client asks, "How can you stand to listen to people like me all day?" or "Why do I get myself into these messes?" The interviewer responds, "I'm here to help. I believe that you have the ability to do something about the messes! Let's try to find out what can be done."

THE SELF-PITIER: The client talks freely about himself but back-peddles into blaming others, martyring himself for others, and excusing himself for not doing things. The client seems not to be making any effort to use the interview to move toward improvement. In externalizing the causes of his problem, the client expects the people around him to change and removes himself from the solution. The interviewer says, "I have the feeling that we are up against something that is stopping us from looking at your problems. Together, let's search further for some solutions."

THE ANXIETY-INCREASER: The client bursts into tears. Talking does not seem to bring relief or reduction of feeling. Instead, emotionality increases and feeds more anxiety reactions. The interviewer comments, "How do you feel now that you have cried?" or "I know you feel bad. Tell me more about it so I can help" or "What do you think would happen if you rode on the subway?" or "How did your feelings about large crowds in department stores begin?" or "What were things like before you became so upset?" or "Talking seems to upset you" or "That feeling is common when a person gets upset."

The labels attached to the client's behaviors just described do not signify that there are distinctive types of clients who exhibit specific behavior. Clients are prone to behave in many different ways. Nevertheless, threads of behavior are usually detected that persist within the interview situation. It is up to the interviewer to detect these recurrent behaviors and to learn how to handle them.

In Example 6-9 the interviewer is accomplishing one or more of the following objectives in each of the sample situations.
• Guiding and involving the client in further exploration of his problem
• Pointing up the client's responsibility for

participation in the interview transaction
- Encouraging the client to examine his feelings about his behavior
- Reassuring the client when necessary but urging the client to move on
- Reinforcing the client's ability to change

Sometimes it is the interviewer who needs to have process limits clarified. Example 6-10 shows behavior that restricts the interviewer's ability to function effectively.

Example 6-10
RESTRICTIONS ON INTERVIEWER'S EFFECTIVENESS

THE COMPETITOR: The trainee wants to perform better than his peers or his teacher or the employed interviewer wants to make points with his supervisor. This hunger to compete frequently results in a defiant approach toward peers, teachers, supervisor, and, worst of all, toward the client.

THE DEFENDER: Because the trainee/interviewer fears losing approval of his teacher/supervisor, he covers up his fears with coping strategies that temporarily conceal his fears of failure. Although these patterns of behavior may reduce the anxiety feelings, these unsatisfying behaviors also dodge the interviewer's problems. As a result, the interviewer pushes his energy toward self-defensive behavior and resents what he considers the teacher/supervisor's infrequent or inadequate praise. Thus the self-involved interviewer diminishes his sensitivity to the client's needs because he is hypersensitive to his own needs.

THE SELF-BLAMER: The trainee feels guilty about using a client as a guinea pig or the employed interviewer feels that someone more experienced could do a better job. Believing themselves incapable, they act incapably. Their performance is consequently lowered—just one more self-fulfilling prophecy.

THE NONLISTENER: The interviewer is so involved in his own problems that he is unable or unwilling to listen to the client. Therefore he misses a great deal of information pertinent to the client's problem and also gives the client a feeling of nonimportance.

THE SPEEDER: The interviewer fails to give the client sufficient time and opportunity to answer a question or to complete a statement. Part of the reason for speeding up may be that the interviewer is convinced that the client's background experiences are so inadequate that the client would have little to offer to increase understanding of his problem. The interviewer diminishes the flow of the client's discussion by a bombardment of talk and explanation, or he may be more direct by remarking, "That's all you have to say about it, isn't it? You're just repeating the same thing from another angle now. Let's talk about . . . ."

THE AVOIDER: Fear of making mistakes stops the interviewer from speaking. Wary of saying the wrong thing the interviewer tends to overlook situations and is quiet and passive when active intervention would be desirable. Insecure about involvement, intrusion, or a possible struggle, the interviewer proceeds overcautiously.

THE OUTLINER: The interviewer is preoccupied with the outline of required information that is part of the intake interview. Therefore he compulsively follows the formal guide, resulting in a disconnected interview of high quantity but low quality. In addition, strict adherence to form reduces the interviewer's ability to listen to what the client is saying.

THE ADVISOR: The interviewer gives advice too readily when the client should be provided with alternatives from which he may select his own action. This does not mean that the client's request for advice should be disregarded. Some direct requests from anxious and dependent clients must be satisfied. Giving advice in this instance is proof of the interviewer's interest in the client and instills client confidence in the interviewer. Later on in the interview transaction the interviewer may offer alternatives rather than direct advice.

The interviewer's personal problems may make the interpersonal process between the client and the interviewer more complex and unsatisfactory. In the preceding discussion a number of ineffective interviewer characteristics are brought to light. The "pointmaker" interview may become defiant in his competitive efforts to get ahead and may also cover up his fears of losing out by lies, distortions of the client's problems, or by excessive self-blame. Any one of these characteristics can blunt the interviewer's ability to listen.

In addition, because of his feelings of unimportance, the interviewer may seek "points" from the client. He rushes headlong to show the client that *he* (the interviewer) *knows*, that *he is wise*. Consequently, the interviewer interrupts and speaks rapidly, disregarding the client's replies. In other situations the interviewer's fear that he may mismanage the interview cuts down his comments to a minimum. Two further obstacles decrease the harmony of the interview transaction. The too-directive interviewer who oozes instructions to the client on the best course of action is similar to the interviewer who is managed by an application form. The interviewer-advisor does not permit the client the opportunity to choose from alternatives; the toe-the-form interviewer does not allow himself an alternative to the rigidity of the words on the intake application.

Handling feelings

The beginning interviewer who has successfully accomplished ways of greeting a client and has begun to be able to set limits is still confronted with the task of handling feelings. Both the release as well as the control of feelings are involved in the interview transaction. In this transaction the interviewer must become aware of both the client's and his own feelings.

Differentiating feelings. The first step in coping with the feelings brought forth during the interview session requires the ability to differentiate feelings. Unfavorable barriers and misdirection result from mistaking the meaning, intensity, or direction of the client's feelings.

Manipulative and expressive feelings. Some clients use feelings as a strategic maneuver to steer the interviewer into compliance with a hidden plan they have devised (manipulative). Other clients attain relief by releasing their feelings (expressive). Releasing feelings is a necessary part of the interview transaction. However, the purposes for the release are also important. The client who uses his feelings as a weapon to urge or embarrass others is repeating destructive behavior that probably has worked outside of the interview. Such a client has probably been using emotional "wheeling and dealing" to manipulate people to do his bidding. The interviewer who gratifies (reinforces) the client's requests during and after the client's emotional maneuvering is simply perpetuating the client's self-defeating behavior.

One of the consequences of the client's manipulation through emotion may be *triangulation*. Triangulation refers to pulling in the interviewer or some other person as an ally against someone else. Sometimes the exploitation or manipulation is accompanied by weeping, by destructive or seductive remarks, or by helpless and hopeless behavior. In each of these behavior patterns the end result is to drag the interviewer into the client's battle on the client's side. Example 6-11 shows the difference between triangulation and the role of a helping ally.

Example 6-11
TO BE OR NOT TO BE TRIANGULATED

MS. S.: *(Beginning to sob quietly.)* Do you have a Kleenex, Doctor? I'm getting overwrought. But . . . but . . . . *(Ms. S. takes the tissue, wipes the tears from her face, and noisily blows her nose.)* He says such cruel things. Just yesterday he called me a bitch. Me! Can you imagine that?

1

DR. M.: Yes, that does sound mean. What happened then?

MS. S.: He banged on the table and asked for his supper. He must have been crazy, thinking I would get it for him after he acted that way.

DR. M.: Hmmm, yes, I see.

MS. S.: Oh, I do feel better, Doctor. You do understand. See how much I am put upon.

Ms. S. goes home to tell her husband that the doctor agrees that he is cruel and mean and does not deserve anything from her.

2

DR. M.: *(Silent.)*

MS. S.: Imagine that. *(Wipes her tears with the back of her hand.)* He's a brute, a beast. He even banged on the table and asked for his supper. He . . . he . . . . *(Wipes her eyes again.)* Doctor, do you have another Kleenex?

DR. M.: From what you say you do seem to be having a difficult time. *(Places the box of tissues on the edge of his desk near Ms. S.)* It would help if you would tell me more about yourself in this situation—how you feel about yourself . . . .

Note the difference between Dr. M.'s first answer and his alternative answer. In Dr. M.'s first response, triangulation is apparent. Ms. S. manipulates Dr. M. to her side in the clash with her husband. Dr. M.'s second response places him in the role of a helping ally, not an ally against Ms. S.'s husband.

In the second answer Dr. M. redirects the client's thinking and feeling to an examination of her role in the conflict with her husband. Dr. M. also notes the persistence of the client's requests for tissues. In the first answer, Ms. S. requests a tissue and Dr. M. automatically hands her one. For the alternative answer Dr. M. places the box of tissues near Ms. S. so that she might remove the tissue herself.

Sometimes requests for tissues or cigarettes or other items are efforts at manipulation of the interviewer through helplessness/dependence. When such requests continue, the interviewer may be catering to dependency needs rather than the client's personal growth.

Tear-wipers and other items should not be refused without considering the immediate needs and functioning level of the client. Offering the client a tissue may serve to assure the client of the interviewer's interest and emphatic understanding of the client's feelings. The interviewer should be aware, however, of the possible manipulative quality of the client's emotional expressions. The more experienced interviewer is able to distinguish with greater assurance between manipulative and expressive feelings.

When the client ventilates his feelings expressively, he achieves relief of his tension. Unlike the manipulative emotional outbursts directed against some person or against the interviewer, the expressive emotional release permits the client to achieve new understandings about his problem. Emotional expression frees the client so that he is better able to do something about what he wants. The effective interviewer assists the client to find a middle ground between release of and control of emotion. The interviewer *encourages* the client who is fearful of yelling at people, showing anger,

and crying, to yell, be angry, and cry. When this client finds it is all right to feel and to express these feelings without losing the interviewer's respect, a great burden of suppression is removed. Clients are encouraged to accept and admit their feelings but not to harm themselves or someone else by acting out their feelings.

The interviewer *discourages* the games some clients play when they assume helpless or "dumb" behavior and speak or cry constantly about being abused or confused. He also discourages the client who attempts to manipulate him with such remarks as, "I feel like crying. He does it. Now why don't you see how wrong he is? I'm right, yes?"

The interviewer is alert to signals that the client is holding back expression of his feelings. Although the client is assured of the right to feel, the interviewer is also alert to the false tranquility of repetitive emotional outbursts. The interviewer should help the client extinguish this inappropriate behavior that brings only temporary relief. The objective of emotional release is obvious. It is to give the client added freedom in using his energy to confront and to resolve his problems. By means of exploring the way a client feels about his problems, these problems take on a new clarity.

Exercise 6-9 focuses on the manipulative and the expressive emotional styles.

Exercise 6-9
MANIPULATIVE AND EXPRESSIVE EMOTIONAL STYLES

This exercise requires a group of four participants and a tape recorder. An interviewer volunteers or is selected and then chooses someone to be the client. Both the interviewer and the client then select "doubles." These doubles will act as extensions of the interviewer's as well as the client's thoughts and feelings. The doubles will not interpret or explain what the interviewer or client says but instead will reflect what they believe the interviewer or client leaves unspoken. The entire interview should take no more than 5 minutes and should be recorded. Select from the interview situations listed at the end of the exercise.

First interview

The interviewer and client and their doubles seat themselves so that the interviewer and the client face each other directly, while the interviewer's and client's doubles are at their sides but slightly to the back of them.

The client and interviewer continue the interview for 5 minutes after the particular situation is described. The interviewer's and client's doubles comment as they see fit, depending on whether they feel that some unspoken thought or unexpressed feeling is omitted. The client and interviewer may use the comments of the doubles to change their procedure, may disagree with them, or may ignore them.

The tape recording is then played back, and for 10 minutes the interviewer and the client as well as their doubles discuss how they felt about themselves during the interview and how the interview might be improved. Some topics to discuss are the following:

1. How are feelings expressed?
2. Describe the feelings expressed. Are the feelings predominantly manipulative or expressive? Are the feelings positive (favorable, pleasant, satisfying), negative (unfavorable, unpleasant, destructive), or ambivalent (partially favorable, sometimes unfavorable)?
3. How do the interviewer and client feel about the comments of their doubles?
4. How should the interviewer change his procedure in order to handle the emotions expressed more effectively?

Second interview

The interviewer and the client as well as their doubles seat themselves as they were for the first interview. A second 5-minute interview is taped, with the doubles commenting as they did before.

After the playback of the second interview, the four participants hold a 10-minute discussion that concentrates on the following questions:

1. What improvements are there in the way the interviewer handled the interview situation?
2. What evidences are there of positive growth for the client?
3. What might the interviewer do to have a more effective interview?
4. What differences are there in the expression and handling of feelings?

Interview situations

1

Katra is a 24-year-old woman dressed in faded blue jeans and a jacket to match. She is wearing a knitted hat, the brim of which floats in scallops over her forehead. In her hand she carries a crushed shopping bag from which she pulls out an application blank.

KATRA: Here it is, my application.

MS. P.: Hello, Miss L. Please be seated. *(Ms. P. takes the application and looks at it.)* I see you have a bachelor's degree in psychology, and . . . .

KATRA: Yes. Hasn't helped a bit. Waste of time. Can't get a job.

2

Grant, 12 years old, walks into Dr. L.'s office slowly. Dr. L. rises to walk toward Grant, but Grant walks past Dr. L.'s extended hand and begins to take large steps around the outer edges of the office. As he walks he holds out his hand and rubs it along on the wall, making a soft swishing sound.

Suddenly he sits down on one of the straight chairs facing Dr. L. Grant slumps down so that his legs extend out far enough to almost touch Dr. L.'s shoes with the soles of his shoes. He places both fists between his legs and stares at Dr. L. Then he begins to scrape the heels of his shoes on the rug and watches the ridges his heel marks make in the rug.

Slowly he raises his hand to his nose and places his index finger into his nostril, moving it back and forth, digging at the interior of the nostril. While he is doing this nose-probing, he stares at Dr. L. and pulls his lips into a thin line.

3

MS. E.: *(Bursts into the interviewer's office and plops down on the long, black leather couch. Looks at the interviewer, Dr. T.)* Why did you do it?

DR. T.: Do it? Do what?

MS. E.: Tell me to talk to him that way.

DR. T.: Tell you? Him?

MS. E.: Yep, last time. I told you about Carl's constant nagging and you said . . . . *(Opens her purse and yanks out a tissue with which she begins dabbing at her eyes.)*

Categories of feelings. In Exercise 6-9 an additional differentiation is added to manipulative and expressive feelings. In the first interview (question number 2), feelings were broadly differentiated into three categories: positive, negative, or ambivalent. Fine (1968) identifies positive feelings as those that are ego constructive (self-constructive). Negative feelings are those that are ego-destructive (self-destructive). Ambivalent feelings are conflicting or contrasting feelings or attitudes about the same thing, event, or person.

In the interview situation the client is afforded an unusual opportunity—the freedom to feel without guilt and without loss of face. The interviewer does not deny the importance of any feeling to the client. In fact, the interviewer supports the client's experience by the psychological closeness of acceptance and empathic understanding. Acceptance does not mean approval or agreement. It means respect, client self-direction, kindness, interest, empathy, equality, and communication.

The interviewer stimulates free-flowing expression so that the client may open new pathways to working through his problems. The interviewer guides the client through his emotional maze, helping him see those feelings that are not fulfilling because they are weapons against himself. These self-destructive (ego-destructive) feelings must be explored so that their threatening power will be diminished. The client eventually asks himself, "What makes me say or feel those horrible things about me? What makes me feel others are okay and I'm not okay? What do I want of me?"

Some clients become agitated when they express their emotions. They feel they reveal too much about themselves. They may even fear they may lose control of their emotions. Loss of emotional control may be of particular concern to the client who prides himself on being cool. Other clients are flustered when they unveil both love and hate about a parent, a spouse, a child, or, in fact, anyone. Such ambivalence (contradictory attitudes) has often been carefully hidden for a long time.

When the client believes that the interviewer is interested in him as a person, he is less likely to leave the interview with an unfinished feeling of "Why did I say it, why did I go so far?" The burden of handling feelings falls on the interviewer. The shocked or surprised interviewer often causes the client to respond more intensely or to "clam up."

Exercise 6-10 offers the opportunity to practice responses to the three categories of feeling.

Exercise 6-10
RESPONDING TO POSITIVE, NEGATIVE, AND AMBIVALENT FEELINGS

The chart on p. 112 presents three expressions of feelings. In the first blank column each trainee should write a reply that would help the client grow in understanding and self-acceptance. In the second blank column the trainees should write the client's response that indicates he understands the interviewer's remarks. After completing the replies, the group discusses them.

Cover-ups for feelings. Culture imposes numerous restrictions as to how, when, and where it is proper to show feelings. These restrictions extend to different requirements for the kind and intensity of feelings for men and for women. Unfortunately these approved and disapproved expressions are not based on biological considerations. As a result, most individuals learn to regulate the expression of their emotions, and the resulting cover-ups are the basis for tension and distress. Sometimes this distress and the anxiety involved are channeled into bodily complaints (psychosomatic symptoms) and even serious illness. Cover-ups may take the form of headaches, ulcers, allergies, or other coping strategies such as defense mechanisms. This does not mean that the physical symptoms are "all in the mind," for the stress of anxiety through chemical disruption may eventually result in organ destruction, and the pain is real. Defense mechanisms also require energy and tend to increase distress. For these reasons, cover-ups may serve as temporary relief but, if continued, they block the way to self-fulfillment.

In the interview the client may conceal his feelings by blocked communication. The client either utters partial thoughts or responses or does not respond at all. At other times the restricted disclosure of his feelings

CLIENT SAYS	INTERVIEWER REPLIES	CLIENT SAYS
I know I should love my baby. But . . . but . . . sometimes I hate her so much it . . . it is frightening.		
I never knew I could feel so beautiful. A glory trip. Spiritual orgasm. Heard—saw the finely etched lines of a leaf swaying back and forth sounding soft sitar sounds. But now, 5 months later, flashback . . . now it hurts. What's happening to me?		
Cut the crap, Doc. Who the hell do you think you are, telling me you understand? Whitey understanding a black man. Shit talk. Nothing. Whatever I do is nothing. Try, try so hard. But nothing.		

is indicated in irrelevant or distorted remarks.

Many factors may restrain the client from open expression. He may feel threatened by the interview situation or may be reluctant at that particular interview moment to share his feelings. Conflicting feelings and confused impressions may make it impossible for the client to find a response he considers appropriate. Even more important is the client's belief that the interviewer cannot or will not understand his feelings.

Culture, the client's personal reactions to the interview situation, and the interviewer's behavior are three factors that may restrict the feeling-message. Of these three factors, the interviewer's behavior is the most relevant since it is the interviewer who must clearly understand the "stop" and "go" signals revealed in the client's behavior. It is also the interviewer who must alter the interview environment so that the client may more successfully free himself from the restrictions on his expressions.

The interviewer must look first at himself to determine whether there is some question, remark, or nonverbal cue of his that discourages the client. If the general pattern of the interview transaction has been favorable and open and then the client changes the nature of his verbal and nonverbal responses, the most immediate cue would be particularly important. On the other hand,

if the client has avoided certain topics and kept to a private world during most of the interview, other difficulties are apparent.

Thus the empathic interviewer who puts himself into the client's shoes will not necessarily open all channels for client expression but will be constantly alert to the client's level of communication and resistance. In the dynamic flow of the interview the distressed client shows his difficulties through blocked communication, and the observant interviewer demonstrates his skills of communication by helping the client break down the barriers.

Exercise 6-11 uncovers some of the problems that may be encountered with cover-ups of feelings.

Exercise 6-11
UNCOVERING FEELINGS

The class is divided into small groups of at least four participants. The groups have 15 minutes to decide which emotional factors influence the clients in the five following examples.

One of the examples is selected. One participant role-plays the client, another participant role-plays the interviewer, and the two remaining participants observe and record the interview session for 5 minutes.

The group discusses the favorable, positive procedures used in the interview to help the client understand his emotional responses. Then the group discusses ways in which the interview approach may improve for 10 minutes.

For 5 minutes the participants role-play the same situation. After the role-playing, the group discusses for 10 minutes any improvement shown and then makes suggestions for further improvement.

1

Ms. L. is having problems with her 15-year-old daughter Emmy. She has come to the high school counselor to find out about Emmy's behavior at school.

MS. L.: We would like to know how Emmy is doing at school. At home she is a young tyrant. We don't know what to do with her any more.

2

The interviewer, Dr. L., has become annoyed with the way in which Mike has been avoiding his questions. Dr. L. is certain he is asking questions that are important to get the information he thinks is necessary.

DR. L.: Now, Mike, I do want to help you. I know you got that dope from someone else. Who was it?

3

The interviewer, Mr. T., is a court-appointed social worker whom Celia has been ordered to visit.

MR. T.: Come now, Celia, are you crying because you were picked up for prostitution or because you can't see your baby?

4

Tom is 16 years old and has been discussing his problems with Ms. M.

TOM: Yep.
MS. M.: Yep? Are you speaking to me, Tom?
TOM: Just thinking. Came out. *(Silence.)*
MS. M.: We were talking about your hands . . . about where they are, or what they are . . . I think you were saying.
TOM: *(Frowns and slumps down in the chair with his hands hanging loosely at his sides.)* Gotta go soon.

5

Grant looks down suddenly and begins tapping his fingers on the desk in front of him. He moves his head slowly from side to side with his eyes downcast. In the 5 minutes of silence he occasionally moves his lips and glances upwards.

Support or identification. The interviewer who is "with it" in the client's scene is able to differentiate feelings and also to help the client "hang loose." To hang loose implies an openness for both the interviewer and the client to explore prevailing hang-ups. When the client or the interviewer faces a problem that is too heavy to manage, he may disregard or even deny the problem, thus draining it, for the moment at least, of its emotional power to trouble him.

The client who rejects a problem regards it as *ego-alien*. In other words, the problem does not belong to him because it does not fit into what he thinks of himself. However, if, through exploring, the client is made aware of the problem's relationship to him, an emotional overcharge may result, leaving him even less able to ponder solutions. Thus the problem becomes *ego-syntonic* and so much a part of the client that he will not or cannot put the problem aside. The client identifies with the problem; he becomes concerned that if he or someone else does anything about the problem, then he is being attacked and might even be emotionally destroyed.

The interviewer may reveal similar responses in his relationship with the client. When the interviewer is too far removed from the client's drama, he will not understand the client's script. The client's problem will be ego-alien. The interviewer may go so far as to reject the client's problem as ridiculous, illogical, or unwarranted. Such an interviewer will not be able to establish rapport or harmonious communication channels.

In contrast to this ego-alien stance, the interviewer may become involved in the client's feelings because they remind him of some of his own emotional struggles. Thus the client's problems become ego-syntonic for the interviewer. The interviewer who wraps himself in the client's problems to the extent that he acts out the client's script is overidentifying with the client.

Dory Previn's haunting lyric (1970) penetrates the depth of feeling resulting from the frustration of overidentification.

i was you

i smiled
your smile
till my mouth
was set
and my face
was tight
and it wasn't right
it was wrong
i was you baby
i was you too long

i said
your words
till my throat
closed up
and i had
no voice
and i had
no choice
but to do your song

i was you baby
i was you too long

i lived
your life
till there was
no me
i was flesh
i was hair
but i wasn't there
it was wrong
i was you baby
i was you too long
and baby baby
the worst thing
to it
is that you let me
do it
so who was weak
and who was strong
for too long baby
i was you

It is not always easy to note the differences between supporting a client and identifying with him. Empathy does not require identification. The "as if" quality of empathy suggests that the interviewer remains himself but feels "as if" he were perceiving the world through the client's eyes. Support is empathic and implies two people—one helping person, one person being helped. Over-identification erases one's own self-identity and substitutes the self-identity of another. The two selves, client and interviewer, merge into one.

Identification may work both ways. The interviewer may find the client's feelings and conflicts so similar that they become ego-syntonic to the interviewer. The client may find the interviewer's clothing, his confidence, and his behavior so appealing that he imitates him in an attempt to be like him. The interviewer loses his value to help when he identifies with the client, but the client profits from the model provided by the interviewer.

In the process of imitating and identifying the client learns new behaviors. These new behaviors are essential to his personal growth. However, the interviewer must be keenly aware of the transition the client must eventually make to his own behavior style. As Dory Previn so aptly disclosed in her lyric "i was you," "i" must not always be "you" or "i" am lost.

The importance of the interviewer's being *with* the client but *not being* the client can-

not be denied. The former is support; the latter is identification.

DEVELOPING OR EXPLORING PHASE

The beginning stage of the interview merges into the middle or developing phase without any obvious point of change. Certain previous concerns become even more apparent at this time and new concerns arise.

Getting into the center of the interview

It may be assumed that the center or core of the interview has been reached when both the client and the interviewer settle into their conversational niches to begin some serious explorations.

Developing objectives

Some objectives for the developing phase are merely expansions of what has gone on in the interview up to this point. The fact-gathering process is justified only if it serves the client's needs for increased understanding. If the opening phase has been successful, the empathic interviewer has already established an acceptant atmosphere in which the client will throw more light on his current problem, be moved to discover whether other problems are involved, discuss ways in which he has tried to cope with his problems, and search for and select from alternative solutions.

Current problem. Up to this moment in the interview the client has been generalizing about his problem. Now it is necessary to obtain more details about the client's concerns as well as to help the client recognize his role in his difficulties. This closer inspection may reveal circumstances that the client has been avoiding in order not to recognize and accept his own involvement and responsibility for his predicament.

The competent interviewer is aware of the double bind that faces the client. The client wants help and yet is fearful that his self-protective behavior may be penetrated. The client therefore holds on tightly to his coping strategies.

The interviewer's skill is vital in smoothing the way for the client through what may become a rough period. It is during this search for details that the client may seek to deny his role in the muddled situation. The client may distort the facts to put himself in a more favorable position or adopt coping devices that apparently reduce the cognitive dissonance (conflicting ideas) yet that continue tension-producing behavior. Thus the client may just move into a slightly different storm area.

Exercise 6-12 offers different approaches to this search for details.

Exercise 6-12
MOVING INTO DETAILS

Each trainee is to complete the interviewer's and client's discussion in the following examples of behavior with three purposes in mind. First, the interviewer encourages the client to recognize himself as one of the sources of the problem. Second, the interviewer reinforces the client's venture from generalities to details. Third, revised and/or new objectives are mutually derived.

After the trainees have finished writing as much as they think is necessary for the five examples, the conclusions to the examples are discussed by the group. The group decides which interviewer responses are effective in attaining detailed client answers and examines the appropriateness of the objectives for she client's presented problem.

1

Mr. D. and his wife are in the office of the marriage counselor because of what they call a "battle of ideas." Mr. D. is certain that he is right. If his wife were not a spendthrift, there would be no problem. Pointing at Mr. D., Ms. D. clinches her teeth and says, "If only you gave me other satisfactions. The only thing I've got—the only pleasure. . . . Buying things makes me feel good." The marriage counselor has heard Ms. D. say this same thing in her two previous sessions and has asked Mr. D. to join the counseling session in order to unravel some of the details.

2

Terry is the spokesman for the group. He rapidly launches into the list of annoyances that the other students had prepared with his help. Terry says, "Here it is. They all revolve around your demands for too much meaningless work for what we want to do as human service workers. We want to find

out more about ourselves. You give us mickey-mouse requirements."

3

The situation has become worse. At least that is what the white members of the club believe. Not only are there special cliques with their own ideas but there is also a sharp split between the white and the black members. Bob comes to the club sponsor to tell him about this split: "We sit together according to pigmentation. Blacks—whites. We—they. Just can't seem to get together."

4

Tom enters the interviewer's office slowly. He wears jeans patched with rainbow colored cloths of different sizes and shapes. His crushed, embroidered, faded blue shirt is loosely stuck into his jeans, which are held together with a large safety pin near the top of the fly. Strung through the loops of his jeans is a bright red macramé belt with a hand-engraved buckle. He sits down on the straight chair with one leg beneath him. On his sockless feet he wears rubbed, ripped moccasins. He says, "Wow! Weary jaunt. promised them I would see you. No need, what for? From place to place, crash where I'm at. Hmmmm, that's it. Where I am— commune, somewhere, nowhere, anywhere. Remorse? Parents? Uptight, bourgeois mentality. I wonder sometimes if they're all wrong? Wonder?"

5

This is Ms. H.'s second visit to the community center's social service department. She is thinking about her last visit as she walks up the steps to the interviewer's office. She mumbles, "All week—what a talk session. More like a sermon. Why am I back here? She doesn't know from a hole in her head. Need someone to tell me how to handle his temper. Blue, yes, blue he gets when he gets mad. He's 4 years old and he twists his father around his little finger. Four! Been winding-up the family since he was 2. Has to stop. Fighting with the baby, fighting with me. It has to stop! Gotta find a way to stop it." Ms. H. enters the social worker's office.

Other problems. The problem that the client first states, the manifest problem, may be the only one the client is prepared to share. This itemization of the obvious evidence of the client's distress may simply be a maneuver to allay the anxiety of what the client fears most. Fear of ridicule, of shame, and of rejection drive the client to conceal his part in the drama he enacts about his relationship with his wife, his son, his friends, or his work. The client becomes a child again, trying to buy favor from the interviewer by proving "what a good boy am I." "If only someone would understand me."

Before the client will freely look at himself within his elaborate system of interrelationships, he must establish a feeling of trust for the interviewer. By exploring the major area of the client's functioning as child, parent, spouse, friend, student, educator, employee, or employer, the client and the interviewer arrive at disclosure of other aspects of the client's living arrangements that are influential in the initially presented problem.

When Ms. H. says that her 4-year-old son Sammy manipulates his father, fights with his sibling (the baby), and has exhibited this behavior since the age of 2 years, an alert interviewer asks, "How does you son 'twist his father' around his finger?" "What do you do when this is happening?" "Tell me about Sammy—about his fights with his baby sister." "And, then what do you do?"

The blueprint for coping with the behavior that has baffled Ms. H., her husband, and her children leads into the circumstances surrounding the behavior, the antecedent events, and the consequent events reinforcing the behavior. (See the discussion of the ABC of behavior on pp. 32-34.)

Details are accumulated to provide the client with an enriched awareness of her role as well as her husband's role in reinforcing Sammy's behavior. Information about Ms. H.'s feeling about her husband is coupled to this awareness of the effect each member of the family has on the other. The inquiry goes on reviewing the interaction between husband and wife. Is there some other behavior of the husband or the wife that is disturbing their relationship and is being duplicated in the parental relationship with their children?

The empathic interviewer is responsive to the client's needs. Together they decide when they have reached a plateau and should stop awhile to look for other directions or have arrived at some solutions to try out. The plateau may indicate an altering of the relationships between client and interviewer, some new complications in the client's life space, or the client's desire to mull over what has occurred. Any number of possibilities may crop up.

Problem coping. Two additional facets are inspected during this developing phase. What is being done about the disturbing situation? How successful are these coping strategies in reducing the disturbance? The answers to these two questions determine alternative solutions. Sometimes the client may be reluctant to talk about how he tackled his problems because of the unsuccessful results. In an effort to avoid criticism the client may become silent or may fabricate some success.

Some approaches promoting the client's unraveling of his distressful behavior might be as follows:
- "You seem to set high standards for yourself."
- "Seems to me you're harder on yourself than you are on anyone else."
- "I'm here to help you, not to criticize you."
- "I wonder what makes you believe you can't do well. Look at how well you handled. . . ."
- "It isn't always possible to do things right the first time."
- "Come now. Here you have the *right* to be wrong. So, if it didn't turn out so well, we'll find another way."
- "I know you're finding it hard to talk but I·can only help when you tell me about things."
- "Hmmmm. Have you noticed how uptight you get when someone says something you don't like?"
- "What has happened between us to close you off now?"

These suggestions are meaningless unless they fit the occasion, the interviewer, and the client. The interviewer who is genuinely interested in the client's self-fulfillment knows that he cannot use "canned" answers.

Alternate solutions. When some degree of clarity is achieved about the branching difficulties of the problem and when the existing coping strategies are apparent, then client and interviewer ask themselves, "What else might be done?" In actuality, clarity, coping, and "what else" are all being discussed to some degree at any one interview moment.

Flexibility is essential in establishing guideposts to new action. Both a reorganization of prior attacks on the problem and some innovative coping methods should be brought forward for the client's scrutiny. Underlying the framework for alternate solutions is a belief in the dignity of the client and his right to make his own decisions about change. The interviewer's role is to help the client figure out what has been going on and to seek other ways of behaving. In some instances the interviewer may make outright suggestions.

Sometimes the interviewer must assure the client that others have experienced similar disturbances and feelings as well as the same lack of progress in finding solutions. The client profits from reports about how others strive to face and deal with their problems. He increases his chances for coping by listening to similar but hitherto unrecognized solutions that he may try with his own difficulties.

In the presence of the accepting, nonpunishing interviewer the client feels free to enlarge on his grievances or to express his hostilities. As interviewing time goes on, the client's fears or anxieties are diminished; he becomes more secure in the interview transaction. More energy is transferred from conflicts and feelings of threat to the positive challenge associated with solution-seeking.

Thus the interviewer avoids the do-gooder attitude that regards the client as a child to be shown the right way. Since there is no one right way, emphasis is placed on the client's initiative and capacity for personal growth. This does not mean, however, that when the client needs information the interviewer refuses.

Through the human relationship of the closed-off, protected social unit of the interview transaction, the client is led to examine his problem, himself, and his interactions with others. When the troubled client is assured that he is worthwhile to someone (the interviewer), he is reinforced by the social approval of the interviewer to attempt new ways of behaving.

Incubation period

Sometimes the wisest way to go forward in the interview is to stand still. The remark "They also serve who only stand and wait" by Milton in the poem "On His Blindness" appropriately fits the periods during which very little client work and progress seem to be made. Often this occurs after the client has spoken about his problem as he sees it. This pause permits the client to integrate the parts of his discussion. The interviewer serves the client's needs best when he stands by supportively but does not push the client into deeper exploration before the incubation period is complete.

The client may begin to say something, veer off to some other topic, and return to what he was saying previously. His thinking appears to be circular, but actually he is moving to some associations with the beginning thought. These associations seem to be drawing him somewhat astray but are meaningful to the client and in the path of his concerns. The client is going through the process of coordinating his thoughts, seeking alternate solutions, and trying to understand what has been going on during the interview.

The incubation period is not the same as inhibiting or blocking of responses. One essential difference is observed. In the incubation period the client does not return immediately to his original statement but, instead, begins an entirely different, often tangential discussion. Blocking shows in behavior that is negative (the client refuses to answer), angry ("Stop getting into my head!"), coy ("Come on, Doc, you don't really want to know that!"), or in some other form of refusal to continue the interview.

In contrast to blocking, incubation offers the client the opportunity to shape his thinking so that he may find a more comfortable way to express a disturbing thought or to bend a little more toward solving his problems. It is a process of organizing and reorganizing on a subvocal level. Incubation may be revealed nonverbally rather than verbally by the furrowed brow, the raised eyebrow, and silence. None of these signals of incubation can be interpreted out of the context of what has gone before and what continues afterward. The effective interviewer does not jump in where the client is not ready to tread.

The puzzled interviewer may wonder, "What do I do or say at a time like this?" If he is really uncertain of what makes the client slow in continuing, he might be honest about it and say, "Frankly, I don't understand what is making you have difficulty in continuing." In the event that the interviewer is able to note a pattern in the client's offshoots of the central concern, the interviewer may comment, "I feel it's somewhat hard for you to go on. Sit and think for awhile if it helps. I don't mind."

Shifting gears

In order to push further into the center of the interview, it is necessary to find some ways to "shift gears" from the beginning of the interview to the middle or developing phase. It may be necessary to go back to what already has been discussed or just to wait until the client gives the interviewer the "go" signal so the gears might be shifted. Four points related to shifting gears will be discussed.

Speed of shift. Transitions from one area to another vary from rapid to slow. Sometimes the transition may be abrupt, as when the interviewer comments, "Now that you have told me your problem, tell me more about yourself—how you feel right now."

Getting the client to talk about himself may be very difficult since he has been accustomed to looking at himself as others see him. He is "no good" because they frown on him as a black man. She is "not wanted" because her father really wanted her to be a boy. He is an "ungrateful son" because he wants to live in a commune. These labels conceal the client's true feelings and communicate a distorted view of himself.

Self-appraisal. The next step, after the client moves from generalities to details, is for the client to see himself as he is, not solely as others see him. When the client says, "I hate being a homosexual, and I don't know what to do about it," the interviewer may reply in one of two ways. He may respond with a statement that prompts the client to continue to look at his behavior as others perceive him, "Then why do you keep on being a homosexual? You say you hate it. You say you can't stand what others say to you." A contrasting comment by the interviewer is, "You say you hate being a homosexual, but you continue. I wonder what it really means to you?"

When the interviewer presses the client to examine what his behavior really means to him, self-exploration becomes the responsibility of the client and the client forges ahead toward a better understanding of himself. The interviewer does not desert the client during this self-probing but is part of the searching team. The client provides the lead, while the interviewer listens sensitively and responds in the client's direction.

At times when the conversation begins to slacken the interviewer may say, "Well, I guess that covers the way you feel about your boss, now let's look some more at what you were saying about your husband." In order to change the topic in this fashion, the interviewer must not get too far ahead nor too far behind what the client is conveying. The drift of the interview still comes from what the client is talking about.

Client's pace. In order to match the client's pace the interviewer must avoid jumping into some point because of his own craving. The interviewer is bored, tired, or annoyed, so he shoves the client into a verbal situation for which the client is unprepared. What the interviewer must do is observe the client's posture, facial expressions, and gestures. On the basis of nonverbal observations plus the client's verbal clues, the interviewer must decide when it is time for the client and the interviewer to indulge in a "pause that refreshes" or when it is time for the interviewer to help the client over a hurdle with some remark such as:

- "Tell me more about it."
- "Tell me how that fits in with what you said before about your. . . ."
- "Tell me your reasons for. . . ."
- "I'd like to hear more about your ideas on. . . ."
- "How do you feel about your relationship with your . . . ?"
- "Anything else?"
- "What does Marcia's refusal to see you mean to you?"
- "Hmmm, tell me where that thinking leads you."
- "How did you come to that conclusion?"

Nonverbal affirmation. The interviewer may affirm his interest in the client by nonverbal cues. He sits forward looking at the client, not above or to the side of the client's head. A nod of the interviewer's head as well as his smile affirms that the interviewer is closely following what the client says and that he anticipates more discussion.

These postures are not mere gimmicks but are means to establish trust and understanding. Shifting gears requires a keen awareness of the meaning of this relationship to both the client and the interviewer and also some further techniques for spurring the client on.

Techniques for spurring the client on

The interviewer brings his humanness to the interview and undertakes a joint effort with the client to accomplish the difficult, active work of mutual understanding. In the process of spurring the client on the interviewer is cautious not to stumble over his own eagerness to make progress. Therefore he must bring his professional responsive skills with him.

Minimal encourages

Ivey (1971) writes about how the interviewer may become "active and involved" in the interview and yet remain client centered so that the directional impetus will be provided by the client's needs. Minimal encourages permit the client to carry most of the conversation while the interviewer mainly listens. By means of sparse verbal and slightly more nonverbal reinforcements, the interviewer is able to provide feedback that sparks the client on to deeper self-exploration. "This skill [minimal encourages] is concerned with helping the client to keep talking once he has started to talk" (Ivey, 1971). In order to perfect this skill the interviewer must learn to be a *good* listener.

The effective interviewer's brief verbal retorts depend on his being plugged in on the client's wavelength. Then the interviewer gives a brief response ("Hmmm," "So," "Oh," "Then?" "And?" "Tell me more," or "Yes, I see,") or repeats a word or a few words that the client just said. Nonverbal responses may be a slow forward movement of the interviewer's shoulders, placing a hand on his chin, or some other characteristic motion of the interviewer.

Such brevity of interview responses tends to be less constraining on the client. The responses' impact on the client urges him to greater self-responsibility and more self-exploration. When the interviewer who adheres to minimal encourages considers it necessary to energize the interview with increased verbal activity, he is less likely to overwhelm the client with a volley of words. Exercise 6-13 offers practice in minimal encourages.

MINIMAL ENCOURAGES

The larger group is arranged into groups of four people. Two people conduct the interview, one person is the observer, and another person is the interpreter. A 5-minute interview is taped, the content of which revolves around any topic the client decides on.

Although the interviewer may freely use nonverbal cues to convey messages of interest and caring, he may respond to the client with only the following words and sounds, "Yes," "So," "Fine," "And," "Oh," "Hmmmm," and "Mmmmm." Even though the interviewer may feel forced and uncomfortable at first, if he really listens to the client, he should become so involved in what the client is saying that he eases up on the mechanical aspect of his responses. The interpreter sits beside the interviewer and puts into a longer sentence what he thinks the interviewer is saying. The client responds to both the interviewer and the interpreter, revealing his feelings about the interviewer's words and sounds and also his reactions to the interpreter's comments. The observer records the nonverbal responses of the interviewer and adds his own inferences about their meanings.

The tape is played back and the four participants discuss the effectiveness of both the interviewer and the interpreter in helping the client understand, feel comfortable, and move forward in his discussion. The observer of the nonverbal cues adds his findings to the discussion. Suggestions are then made for improvement of the interview.

Another 5-minute interview is taped, with this difference—the interpreter's role changes; he records the words and sounds the interviewer is making and also notes changes in volume or tone of the interviewer's responses. The client continues with his discussion, relying only on the interviewer's verbal cues. The observer continues to record nonverbal cues.

After the playback of the tape, another discussion is held with an additional evaluation of the changes made and the degree of improvement. The group considers how the interview may be further improved.

Eductive techniques

Just as the client reinforces certain interviewer responses, the interviewer encourages (positive reinforcements) or discourages (negative reinforcements) the client's responses. The interviewer trains himself, however, not to interrupt, to avoid unnecessary advice-giving, and to be frugal with words and expressive with his body and gestures. The objective is to open the way for the client to ease into talking about himself and whatever is troubling him. The eductive technique permits the client to speak of his concerns without feeling obligated to do so.

Talking about himself is a hard job, probably the most difficult work the client has ever done. The courage to honestly talk about one's problems and the behavior associated with these problems radiates from the interviewer's empathic understanding.

As the client is in the process of reporting his concerns, the interviewer's face-to-face presence provides a sounding board as well as a protective shield. Eductive means to lead forth or to develop. Thus the interviewer's nonpossessive warmth coupled with his genuineness promotes the client's feelings of self-worth and trust. Trust, in turn, assures the client that what he says during the interview will not be used as a weapon to threaten him.

In addition, the client is assured that *he* is the one who selects what he is comfortable in communicating. There is no obligation to dredge up the past for the history of present behavior. The client invests his energy in the present and the interviewer maximizes the meaning of the reality elements of the present by means of the transaction relationship established.

The success of eductive techniques is dependent on the shared learning experience established between interviewer and client. The process of recounting events brings the problems directly to the client's attention. The interviewer becomes the source of reinforcement and the model of learning for the client coping with his problems.

Controlled nondirection

A Chinese proverb typifies what Kahn and Cannell (1957) call controlled nondirection: "You give a man a fish, he has one meal; you teach him to fish, he can feed himself for the rest of his life." In the context of the relationship between the interviewer and the client, this proverb affirms the idea that the interviewer's role is *not to do* for the client but to reinforce the client in learning *how to*

do for himself. This idea incorporates both client direction and the nurturing of this client direction by the interviewer. The client absorbs some of this learning from attending to his problems during the interview situation, some from the prompts or reinforcements of the interviewer, but most from the acceptant atmosphere of the interview situation.

The process of enlarging on and clarifying the client's problems is lubricated by the interviewer's questions or statements. The controlled nondirection arises from the nature of the interviewer's questions (control) and from the shifting of the responsibility for information and searching to the client (nondirection). Minimal encourages as well as the eductive technique follow the same line of thought—client direction within an interviewer framework.

The purpose of any interview is the stimulation of communication. The important consideration is to provide this stimulation without narrowing or modifying the meaning of the client's responses. Both the kinds of interviewer questions and/or statements and the manner in which the interviewer words these quests for information influence whether the client is given the fish (the answer and completion) or is taught how to fish (to find his own answers).

Listening

There is no more inclusive property of the effective interview than listening. Listening and observing form the center around which all other characteristics of the facilitative interview revolve. In spite of its vital impact on the success of the interview, interviewers too frequently fall short of skills essential to listening. Facility in using minimal encourages, eductive technique, and controlled nondirection also depend on these dual abilities of listening and observing.

Several points pertain to the skill of listening.

Characteristics of an effective listener. It has been said that an individual may hear, but does not listen; he looks, but does not see. Listening and seeing require concentration in the here-and-now. Effective listening is built on the cooperation of both the speaker and the listener in advancing toward better understanding. Intelligence as well as sharp vision and hearing support the abilities of seeing and listening but are sterile skills without the communication of empathy. An efficient listener is actively engaged in receiving, recording, and decoding (getting the meaning of) the client's message. The empathic listener develops large eyes, big ears, a small mouth, and positive actions that reveal his interest and acceptance of the client.

Listening has become even less of a skill with the coming of TV in the 1940's. The "tube" reinforced passive behavior and the young grew up on a diet of pictured words garnished with commercial snacks. TV's instant exposure to events does not demand much of the viewer. Pushing buttons and turning dials are often the extent of the experiencing involved.

"Tube plopping" has at least two unfortunate effects. It undermines the process of learning to respond on the basis of listening, and it detracts from the individual's feedback to his own words. Learning to listen to others begins with listening to oneself. The process leading to the achievement of the listening skill encircles input (speaking), response (as a listener), and self-feedback (listening to oneself). Communication may break down at any of these stages. The growing child may not have had ample opportunity to speak, to listen, or to respond because other people were absent or uninterested.

Since listening is the bridge between hearing and understanding, it is imperative that the interviewer attend equally to verbal and nonverbal cues. Competent listening presupposes a receptive state in which the client's glances, muscular twitchings, gestures, and other bodily reactions as well as his tone of voice, his pauses, and his words are noted.

The nonverbal cues must be tied together with what is said out loud. The interviewer must ask himself, "Do the client's words agree with what he exhibits in his nonverbal behavior?" Essentially the question becomes, "Is the client's behavior congruent?"

Attending behavior. When the primary characteristics of the capable listener are put together, they include accurate observation of both verbal and nonverbal cues, active concentration on the spoken words, and caution in interpreting the client's meaning. The efficient listener *stops* to listen. He is occupied with the client, not preoccupied

with his own thoughts. Certain activities support the notion that the interviewer is with the client. Ivey (1971) calls these activities attending behavior.

When attention is paid to the client, the interviewer expresses his attending behavior both verbally and nonverbally. The interviewer's choice of words tells the client that the interviewer is listening to what the client says. What the interviewer says falls into line with the client's needs for clarification, not the interviewer's oversolicitude or curiosity. The alert interviewer avoids the frequent use of "I" and is cautious about the inclusion of the client into "we" until the client indicates his readiness and willingness to become part of the social dyad (a group consisting of two people with some degree of organized interpersonal relationship).

The nonverbal cues come from the interviewer's body movements, facial expressions, and tone of voice. The comfortable interviewer is "physically relaxed and is seated with natural posture" (Ivey, 1971). Some interviewers who are attending closely cross their legs; others keep their feet flat on the ground as if ready to spring forward. Crossed or uncrossed legs are part of the language of of attention that the interviewer learns. Leaning slightly forward and moving the body slowly backward and then toward the client accent certain points that the client makes and reassure the client of the interviewer's attentiveness.

The inexperienced interviewer may attempt to appear attentive but will reveal his discomfort and even fear by his rigid, upright position, which is often aggravated by the closed-in look of his crossed legs, folded arms, and tight jaw. In order to unlock himself the fearful interviewer may need to learn how to relax. When an individual relaxes, he lets his body unwind. The cat is a wonderful model to follow. When the cat awakens, he stretches his back into an arch, extends his legs, and opens his jaws wide. Slowly he blows himself up. Then the air slowly comes out. The cat now hangs loose. Exercise 6-14 concerns body unwinding.

Exercise 6-14
BODY UNWINDING

Each trainee sits as upright as he can on a straight-backed chair. He pushes his shoulders back and shuts his eyes. The trainee feels the points of contact with the chair and concentrates on what he does as he twists around to be more comfortable. He twists, stretches, sighs, and yawns as he pleases. The trainee should feel himself stretching the muscles of the various parts of his body as he plans the steps in his own slow unwinding.

Some people prefer inching up their loosening behavior from toes to head. Others are more comfortable starting with the muscles around their eyes, ears, and mouth. One should do whatever comes naturally. Unwinding in the presence of an observer offers feedback about the *before* and *after* look of tension and relaxation.

A rigid body keeps sending messages to which the individual must attend. The relaxed person does not have to squirm and shift his position because of muscle tension but can attend to the other person and not so much to himself. He can maintain an easy eye contact that is neither a stare nor a glare. This does not mean that the interviewer must continually look at the client. A shifting (moving) glance—as long as it is not shifty (tricky or deceitful)—is desirable.

Eye contact is often used to reinforce the client to go on with a particular topic (by looking at the client) or to shape the talk behavior away from a blind-alley discussion (by looking away from the client). This discriminative use of eye contact assists the client in exploring his problems and in finding solutions. The interviewer must learn to interact visually so that his eye contact not only shows the client his interest but also reinforces the client's personal growth.

Restful, comforting eye contact comes easy to some people but must be learned and practiced by others. The deliberate learning of the skill of eye contact may make the learner feel artificial at first but, with practice, the approach becomes automatic and satisfying.

Contact between people is built on being on eye level with them. People must be close enough, must face one another, and must be at similar heights in order to be able to communicate successfully. Exercises 6-15 and 6-16 point out the influences of physical distance and of height in the ability to maintain eye contact. Exercise 6-17 shows more directly the importance of eye contact.

Exercise 6-15
LISTENING AT A DISTANCE

Trainees choose partners and seat themselves at opposite corners of the room. The partners talk to each other about any topic for 2 minutes. Then they discuss for 5 minutes how they both feel. Did they look at each other? Did they feel comfortable? How did their voices sound?

The same procedure is repeated two or three more times, depending on the size of the room. Each time the partners move a little closer, talk for 2 minutes, and then discuss the same questions as before. They note the differences in responses that the closer distance makes.

Finally, the partners are close enough so that they are face to face, with their knees almost or actually touching, whichever is more comfortable. They discuss the same questions. Does physical distance make a difference in the ability to effectively communicate?

Exercise 6-16
A CHILD'S VIEW OF LISTENING

Each person should have the chance to be the "low" one in this exercise. One person is seated in a chair and another person sits on the floor in front of him. The person in the chair looks down at the person on the floor and the "low" person returns the look. They talk for about 1 minute. Any longer than that will probably make the neck of the person on the floor uncomfortable.

After talking together in these positions, the two get into more comfortable positions and for 10 minutes discuss how they both felt about their "high" and "low" positions. How did their bodies feel? Were they able to attend to what they were saying? Do they feel differently now that they are seated more comfortably?

After this discussion, the two should switch positions and the entire procedure is repeated.

Exercise 6-17
THE EYES HAVE IT

For 5 minutes two people sit directly opposite each other close enough that they can touch. They talk about what they think the characteristics of an effective interviewer should be. During this conversation both people should be looking at something or someone *but not at each other*.

During the next 5 minutes they talk about how they felt during the first 5 minutes. During this period only one person looks at the other. The other person continues to do what he did before—look at something or someone but not at his partner.

For the last 5 minutes the two persons talk about the differences between the first and second 5-minute periods. Both participants look at one another during this conversation and report on how this feels. Does eye contact make a difference?

Silence. Attending with words, with body, and with eyes is the foundation for feeling more comfortable with the sounds of silence. The Eastern cultures grant room for the silence of meditation. In the Western cultures, however, silence bears the mark of incompetence. Hence in Western cultures, people without a flow of words (the gift of gab) learn to be uncomfortable if they do not know what to say. They treat silence as if it were a vacuum without meaning.

There is no question that silence should be valued more than it is. Silence should not be dressed up as a threat, a failure, or a breach of etiquette. Instead of an absence of communication, silence should be a specific form of communication that is understood in the context of the relationship between the participants.

The interviewer needs to revise his thinking if he considers the client's silence as his fault. Few beginning interviewers can accept silence as a positive form of communication. Instead, they consider silence destructive, indicative that the client is holding back and therefore resistant or unmotivated. Consequently, the interviewer blames himself for being incapable or too gruff. The client must be assured that he will not be pressured into discussing subjects he is not ready to do. The compulsion to talk often stems from the fear of silence. Talking to hear oneself talk becomes the way in which the individual relates to his companions and to the interviewer.

The essential fact is that the client and the interviewer should be aware of their feelings about silence during the interview; then they are able to appreciate the silence and to handle it more capably. The interviewer must hear the client's silence as well as his words.

The interviewer needs to make allowances for client variations in talking style. Some clients speak from the moment they enter the office. Others burst into speaking, stop, and then begin again. Still others say very little, mumble their words, or do not say anything. Whatever the style the effective interviewer learns to listen to the language of words and of the body.

Beginning interviewers feel they must earn their bread. They think that when pauses are long they are not doing enough for the client. This may be so. They should not be doing things *for* the client anyway. The appropriate mix of interviewer listening, pauses, and responses is one of the most difficult interviewer behaviors to learn. To evaluate the significance of a pause the interviewer must be mindful of the timing and frequency of the silence and whether the pause was initiated by the client or by the interviewer.

There is no universal rule about how long is too long for a period of silence. Each interview and each client has to be taken on its own merit. Silence is worthwhile only as long as it is communicating something or serving some function and is not frightening the client. Recurring silence should be carefully examined by the interviewer. The interviewer should explore what he is doing and saying and what is happening between him and the client.

Silence in the early part of the interview is more apt to reflect embarrassment, resistance, or the client's fear of what the interviewer is thinking. As the interview progresses, silence gradually becomes more supportive and a medium for emotional expression and thought.

An effective rule to follow is to let the client assume responsibility for going on if he initiated the original pause. This does not mean that the interviewer sits back and awaits the client's bidding. It is important that the interviewer alert himself to those situations in which the client needs support over some rough places. The interviewer's nonverbal behavior or his brief comments avoid interference with the client's silence but prove the interviewer's understanding that the client is not ready to face his problems verbally. The objective is to encourage the client to take on the responsibility for both his pauses and his words when he is ready.

The client must discover that he can be silent and still be liked. When the interviewer accepts and respects the client's silence, the less articulate client gains a feeling of worth and acquires a feeling of acceptance for what he is, a quiet person.

There is value in the client spending time with someone who understands and who has faith in his ability to solve his problems. This value is apparent even when little is said. Client self-awareness does not hinge solely on verbal exploration. What is even more important is the client's feeling that the interviewer accepts him as he is and that the interviewer does not look down on the client for what the client considers his failures.

The client often uses periods of silence to delve deeply into his feelings, to struggle with alternative courses of action, and to weigh a decision. Often the interviewer senses that the client is rushing too fast or pushing too hard. Then the interviewer may suggest a slowing down to allow longer "think time," longer periods of silence. The interviewer says in effect, "We are not in a hurry; take it easy."

During later interviews or the latter part of a single interview the interviewer's silence tends to have a calming effect on the client. The silence technique does not signify that the interviewer is passive or uncommunicative. The interviewer must learn to talk for the right amount at the right time. The way in which the interviewer moves into the client's silence and the way in which he moves in his chair have bearing on how the client feels about silence. Thus silence becomes a bore, a burden, or an opportunity for getting together.

Silence is easily misunderstood. The same may be said for even the simplest communication. To pause for refueling after an idea is completed is important. To pause because of negative feelings is not as productive.

Heavy silence is unproductive when it represents conformity, discomfort, confusion, or hostility. Often the client is misguided into silence by means of unsuitable timing or reinforcement. Speaking in low, almost inaudible tones is a form of communication that resembles silence.

There are other reasons for silence, however, that make pauses valuable in the progress of the interview. "Headwork" silence is a period when the client or the interviewer quietly organizes and summarizes his thoughts and creates some solutions. Finally, there is a peculiar kind of silence that shows itself by overtalkativeness. The client or the interviewer runs away from silence by chattering. Example 6-12 goes into more explanation of bases for silence.

Example 6-12
BASES FOR SILENCE
TOE-THE-MARK SILENCE

Even though it seems unbelievable in the contemporary life scene that some parents should still insist that children should be seen and not heard, it is not as unusual as some may think. The requirements for obedience during the individual's development may have been subtle. Parents may have talked endlessly and the child was unable to pursue a thought or say a word. On the other hand, the restrictive communication style of the home may have imposed a lock with no key on the child's mouth.

The client reveals his obedient wait-until-you're-addressed approach to conversation by his silence that is broken only when the interviewer asks a question or makes a comment. The client does not talk until he knows what the interviewer wants him to talk about. If the interviewer remains quiet for any length of time with this toe-the-mark client, the client feels that his talking is prohibited.

How the interviewer handles it

The client must be assured that he has the right to be silent. With the obedient client, the interviewer must take a further step. This client must be convinced that he not only has the right to be silent but he also has the right to speak. Some appropriate interviewer comments might be:

- "I'm interested in what you are thinking. It would help us decide on some directions for you to take."
- "I've told you how I feel about the complicated situation in your home. Now, how about your feelings?"
- "Whenever you're ready, you can tell me your ideas on the subject."
- "There must be lots going on. I wonder if you are ready to share it with me."

AWKWARD SILENCE

The awkward silence is an uncomfortable pause that comes when one topic ends and the client is searching for what to discuss next. It also happens as the interviewer and client move toward the termination of the interview session and one or both of them begin to notice an empty space in their word exchange.

At other times the client may flounder because he doesn't know how to begin or doesn't want to begin talking about his problems. He has been rapping casually and then finds a lull in the conversation. This turning point makes the pause awkward because the client does not know what to say as a transition to the more serious concerns he has been avoiding.

How the interviewer handles it

- "I feel as if both of us have nothing more to say about this point."
- "Perhaps we have said all we are going to say today."
- "It seems to me that you are looking at me as if to say, 'Where do we go from here?'"
- "It's sort of hard to get down to serious business."

HOT-SEAT SILENCE

The unpleasant feeling associated with the hot seat of anxiety is part of the awkward silence. The client is often fearful and feels isolated, so he "clams up." Yet the client would be even more fearful if the interviewer were also silent. The interviewer's silence would make the client feel rejected. In order to avert the threat of the interviewer's silence the client may become talkative or fall deeper into silence. In this way the client rejects the interviewer before he himself can feel rejected.

Silence serves an important function for the client who is trying to avoid responsibility for and consequences of verbal communication. By not talking, the client avoids what he considers to be inevitable, namely, unkind remarks and a feeling of being inferior.

Painful feelings may tie the client into another kind of protective shell. Although the desperate client may want to express his feelings, he is not ready to hear the words out loud. The observant interviewer quickly picks up the muscular tension around the eyes and the mouth that signals the distressing experience.

There are other clients who are often disregarded. These are the clients who have difficulty in expressing themselves either because of language barriers, a speech difficulty, or because they have (or think they have) difficulty in putting their thoughts into words. When the interviewer assists this client to break his barrier of silence and attempt to reveal his feelings, the client experiences his first success in self-expression.

How the interviewer handles it

All clients need acceptance and support from the interviewer; the anxious person needs it the most. The warmth communicated by the interviewer's hand on the client's shoulders or hand is a uniting of forces. The interviewer assures the client, "It is all right to be silent. I am with you and will share this quiet period with you until you think of something you want to say."

If the client feels withdrawn and isolated, the interviewer begins the conversation with some information he knows about the client or some item likely to ease into verbal exchange. The interviewer may say, "Up until now, we've been talking about how your father feels about you. Now, how about your feelings for your father?" Of course, the interviewer is cautious about what he says and just when he says it to prevent intensified client withdrawal. Other interviewer comments are the following:

- "It's okay if you want to wait until the words come."
- "Hard to say what you want . . . isn't it? Give me a hint of where you're at. I'll help you find the words."
- "It's okay to say whatever you like. Begin wherever you want."

JUMBLED SILENCE

The client may say something that confuses the interviewer, or the interviewer may say something that confuses the client. In some instances the extended pause that follows a client-interviewer interchange signifies that both the client and the interviewer are confused.

How the interviewer handles it

- "I want to help you. I need to find out more about what you think and feel about this. Let's try again."
- "What I said just now about . . . seems to have confused you."
- "What I meant was. . . ." (Interviewer rephrases his previous statement.)
- "I really don't know where we're at. Let's try to unravel this together."

FOOT-DRAGGING SILENCE

The client becomes reluctant to speak and feels hostile for any of several reasons. He may resent the interviewer's probing and uses the silent treatment as a form of defiance. On the other hand, the client may be asserting what he considers to be his individuality and freedom because he is opposed to the authority figure the interviewer represents. Therefore the client rejects the interviewer through his silence.

On other occasions the client is neither interested in the interviewer nor does he think highly of the interviewer's ability. So the client clamps his mouth and shows his annoyance by means of his antagonistic silence. Passive defiance (not talking) may be the only way the client has learned to fight back when frustrated or angered.

Resistant silence results frequently when someone sends or brings the client to see the interviewer. At first the client may speak with rude words and curt phrases. Then he envelopes himself in a protective silence. In similar fashion the foot-dragging silence of a client may arise from his reluctance to reveal what comes next in his story.

Even more serious for the angry client are his fears that he will not be able to control his aggressive, destructive impulses without the cover of silence. His wordless protest and the passive defiance of silence fight back the overwhelming rage.

How the interviewer handles it

Unless the interviewer views the hostile silence as a challenge rather than as a threat, he will find the client's silent behavior too difficult to handle. It is important that the interviewer avoid a response that shows that he feels personally attacked. Instead, the interviewer accepts the client's form of protest and says something like the following:

- "It seems to me that you are reluctant to talk about. . . ."
- "I don't mind the silence. I do feel you are resenting me in some way. Wish you could tell me about it so we can find out what to do about it."
- "I can wait, but if there is something you're feeling, let's get it out. Let's look at it together."
- "I don't feel that either of us is particularly comfortable with this silence."
- "I don't know what to make of this silence, do you?"
- "You don't feel like talking just now, do you?"
- "Last time we got into a long silence like this you said it was something I had done. How about this time?"
- "Is your silence connected with anything I've said?"
- "Who used to give you the silent treatment?"

RETREAT SILENCE

This form of silence usually happens when both the interviewer and the client speak at the same time. Then both are silent, each waiting for the other to continue what they began to say. This silence is usually much easier to remedy.

How the interviewer handles it

- "Sorry I got in the way. Go ahead."
- "What were you saying?"
- The interviewer smiles and nods encouragingly.

REINFORCED SILENCE

Clients are reinforced to continue or to stop their conversation or their silence by what the interviewer does in response to the client's behavior. Sometimes the client utilizes his silence to reinforce the interviewer's increased conversational activity. At other times the overpowering interviewer may make a silent victim of the client by shaping him to hold back. As a result, the client feels uninvolved in the interview and does not recognize his role in the process of personal growth. The client often becomes even more upset by his deficiencies and more afraid of the interviewer's prestige and power.

How the interviewer handles it

The essential issue is the interviewer's awareness of his power to steer the client. If he finds himself indulging in too much or too little talking, he must pull back to examine whether he really understands the client and what is happening. The interviewer must constantly keep before him the thought that his helping (reinforcement) is for the client's personal growth, not for the interviewer's glory.

INTERVIEWER'S SILENCE

Occasionally the interviewer is working through some thoughts and becomes silent. Fortunately this may be just what some client needs. The interviewer's silence exerts pressure on the client to talk and thus leads the client into greater exploration of his problems. However, another more insecure and/or more hostile client may find the interviewer's silence threatening.

How the interviewer handles it

- "I just have nothing to say right now . . . I'm with you . . . I'm here . . . I'm listening . . . please go on."
- "Truthfully, I don't know why I have so little to say today. Played out, I guess. I am with you though. Do go on."

PSEUDO-SILENCE

An interesting kind of silence develops when either the client or the interviewer speaks so quietly that the other must strain to hear. Whatever the reason may be for this whispering, it hampers communication and may give an impression of disinterest. It is often very difficult to encourage the quiet-speaking individual to increase the volume of his voice. Unless something can be done about it, the receiver of the quiet communication finds the leaning forward a back-breaking exercise in futility.

How the interviewer handles it

- "Sometimes I wonder. Do you really want me to hear what you are saying?"
- "Here goes, turning up the volume button."
- Interviewer says something in a very low voice. "See, it's difficult to hear me. I'm having the same difficulty with your speech."
- The interviewer sits back in his chair.

"You see, it's something I can't do, sit forward to catch your words. You'll just have to speak louder."

SORTING-OUT SILENCE

A sorting-out silence is "thinking silence," which is reflective and productive. The client or the interviewer pauses for a few seconds, during which there is a sifting through and organization of thoughts and feelings to decide on what should be said next. During the client's silence the interviewer may join him in meditation or may use this time to reassure, to inform, or to explain according to the client's request.

Thinking silence is both absorbing and nourishing for the interview participants. There is a cozy feeling of empathy. The client may have just finished talking about an unusual, wonderful event or a frightening, tragic experience. Mutual silence of client and interviewer follows this client disclosure, during which both people need time to recover from the great joy or the emotional shock.

This deliberate, shared silence serves another purpose. It permits the interviewer to listen with a "third ear." Listening with a third ear is the skill of reaching usually locked-out sounds and sights. It means that an individual listens to his own thoughts and feelings and becomes more keenly aware of his sensory impressions.

The interviewer who listens with a third ear is using his thought speed to put together what the client has communicated with his words as well as with his motions. Listening with the third ear obliges the interviewer to keep his antenna up to catch what is happening during conversation as well as during silence. He hears a client's question that appears simple and direct on the surface—"Where do you live?"—and he determines just what information the client is seeking. For one client it may be simply an address or a geographical location; for another, it may be an inquiry about the interviewer's philosophy of life. Thus the interviewer becomes more aware of the client's world by means of his third ear.

How the interviewer handles it

As in all periods of the client's silence, the interviewer is courteous and respects the client's inclination. He does not impatiently interrupt the pause and in the process shut out the client's train of thought. The interviewer responds with silence so that the client has the opportunity to pull his thoughts together for the purpose of answering the interviewer's question or in order to begin a new topic.

Meanwhile the interviewer watches for signals that indicate that the client is reaching out for help in starting anew.

- "There seems to be a great deal going on. I'm ready to participate if you're ready to have me."
- "It must have been an experience."
- "I'm with you, go on. I'll wait if you're not ready to go on."

Nonverbal gestures from the interviewer affirm that the interviewer will not hurry the client into talking. Sometimes this may mean that for this particular session no words are exchanged, just companionship.

PREGNANT SILENCE

The pregnant silence frequently follows the sorting-out silence. Quiet "headwork" continues incubating a creative idea. Interwoven in this creative silence is a feeling of fulfillment. The client discovers solutions. He may not feel that he wants to talk about his brainchild immediately, but the interviewer notes the reduced tension and often the smile of understanding and satisfaction.

How the interviewer handles it

- "You look as if you've found something. I would be pleased if you would share it with me."
- "Wow! You do seem to feel better. What's it all about?"
- "Whenever you feel like sharing your thoughts, I'm here to share them with you."
- "Tell me about it."

ANTISILENCE

Talking too much is often just as annoying as long periods of silence. The chattering client may believe that to remain silent is impolite, unkind, or even snobbish. The interviewer also may find silence hard to bear. He may feel that he is at fault and that he immediately must remedy the silent situation. He pounces on the client and pushes him into talking.

Making conversation creates a wall of words to hide behind. With perpetual words, the client reveals less of his thoughts, wishes, and feelings to the interviewer and, perhaps, himself. Overtalkativeness keeps people at a distance just as effectively as silence does.

How the interviewer handles it

- "Too fast, too fast. Can't follow what you are saying."
- "Let's stop a minute and think this through."
- "Whoops. Lost you there. Back up a minute. What did you just say?"
- "You really know how to explain about. . . . Would you retrace your steps a bit? Tell me about. . . ."
- "How did you feel about the . . . you just mentioned?"

Silence merits this lengthy discussion because it is so misunderstood, misused, and yet required for the effective interview.

Edward H. Richards accurately described the impact of silence in his poem "The Wise Old Owl."

> The wise old owl sat on an oak,
> The more he saw the less he spoke;
> The less he spoke the more he heard;
> Why aren't we like that wise old bird?

This is not a plea for a silent interview but rather an appeal for a more silent interviewer. Exercises 6-18 to 6-20 should help develop an increased ability to listen to silence.

Exercise 6-18
HOW MUCH DO I LIKE YOU?

Person A selects a partner and sits down 5 feet away, opposite to him. The partner, person B, selects two observers who will write down all gestures, body movements, and facial expressions that the two partners make.

For 5 minutes person A tries to tell his partner how he feels about him without using any words. Person B jots down what he thinks person A showed him about how much he liked or disliked him. Then the roles are reversed and person B tries to tell person A, nonverbally, how he feels about him for 5 minutes. After this, person A notes what he believes person B told him about his feelings of "liking" and "disliking."

After person A and person B have completed their 5-minute sessions, all four members of the group discuss their notes and verify whether the silent messages were interpreted accurately. Person A also talks about what signals encouraged him to choose person B as his partner, and person B explains what induced him to select the two observers. These discussions will reveal some additional nonverbal cues.

Exercise 6-19
HOW WELL DO I READ YOUR SILENCE?

Four participants are required for this exercise. Two are the observers, who will carefully write down what they see and hear. The third member of the group is the client and the fourth one is the interviewer.

The client carries on a 5-minute conversation, while the interviewer replies only by nonverbal messages. Then for 10 minutes group members discuss what they saw, heard, and how they felt about what happened during the interview. The client and interviewer reverse roles and repeat the exercise.

Exercise 6-20
COPING WITH THE SILENT CLIENT

Four participants are required. One participant keeps detailed notes of what the client is saying and doing. Another participant notes what the interviewer says or does.

The client speaks for approximately 1 minute about any problem he chooses to relate; then he becomes silent until the interviewer encourages him to talk. The client does not go on unless he feels that he wants to do so. The interviewer respects the client's right to be silent and yet proceeds to restore verbal communication when he deems it advisable. Then the four participants discuss the following four questions.

1. How did the client feel about the interviewer, and how did the interviewer feel about the client?
2. What did the two observers note that the client and the interviewer were saying or doing? What messages were they sending?
3. What successful procedures did the interviewer use to show his respect for the client's silence and to help him begin to speak?
4. How may the interview be improved?

After discussing the four points, the entire exercise is repeated. The same questions are considered after the second interview, with the addition of how the second interview showed improvement.

Silence clings around the fearful person and gives comfort to the composed person. Occasional periods of silence make the conversation between client and interviewer more meaningful. When words fail, silence can help settle distress. The client who is unable to put his problems into words may utilize the sounds of silence more frequently; these sounds of silence will be detected in his facial expression, gestures, and body. The interviewer assists (reinforces) the client to eventually end his silence.

Speaking alone or with another. Moving from silence to sound is not merely a matter of opening one's mouth to spill forth words. Although it may seem contradictory, words sometimes isolate people even more than silence. Human relationships have many levels of communication that range from speaking alone to speaking together. There are five different forms of speaking— the soliloquy, the monologue, the duologue, the duelogue, and the dialogue.

The *soliloquy* is a form of solitary speaking in which the individual talks aloud but to himself. Infants soliloquize when they babble sounds. They hear their own sounds and learn to use certain sounds and certain combinations of sounds later by means of the reinforcements of the significant people around them as well as by the self-reinforcements of their own sounds.

Shakespeare used the soliloquy often in his plays. Hamlet's six soliloquies are similar to free association in which he expresses his feeling of helplessness and disgust with the smell of rottenness and corruption around him. At first Hamlet speaks out loud to himself and does not act. Afterward, Hamlet's soliloquy spurs him on to action.

Soliloquizing may occur during an interview. If the client rambles on about his concerns, this may be valuable for him in order to put his thoughts in order. Instead of silent contemplation, the client sticks his thoughts together out loud. On the other hand, if the interviewer gushes forth a soliloquy, this has little value for the client. The client needs to be involved in the communication process.

The *monologue* is a little different than the soliloquy. The monologue *demands* an audience; the soliloquy *may* have others who happen to be listening. The monologue monopolizes a conversation. One person makes a long speech with little interest in receiving conversational feedback. The interviewer who carries on a monologue clicks away the precious minutes of the client's time unmindful of the client's needs.

Another form of a monologue occurs in the *duologue*. In this communication style, both the interviewer and the client carry on separate monologues. It is as if two television sets in the same room are facing each other. Obviously both persons in this so-called conversation are speaking as if they were alone, although they happen to occupy the same room.

Sometimes this duologue is argumentative and then it becomes a *duelogue*. Each combatant arms himself with word weapons and carries on a verbal duel. Unfortunately for the client the contest is frequently frustrating and destructive since the issues are not considered. Instead, the weaknesses of the participants are attacked. An empathic interviewer avoids such word-fighting encounters.

The *dialogue* contrasts with these one-sided speeches. The dialogue takes into account both the sender and the receiver. The client and the interviewer form a circle of communication, a transaction, in which they openly talk together, seeking the harmony of mutual understanding. In a dialogue the interviewer is alert to the client's leads and encourages the client with a minimal amount of words and direction.

Exercise 6-21 directs attention to the differences among the soliloquy, the monologue, the duologue, the duelogue, and the dialogue.

Exercise 6-21
SPEAKING ALONE OR WITH ANOTHER

Read the five communications and decide which is a soliloquy, a monologue, a duologue, a duelogue, or a dialogue. Then rewrite them so that they are all dialogues. After you have completed the exercise, discuss your answers and revisions with the group.

1

PETER: Hah! Caught ya signifying again. How the hell can you live with yourself, spying like that?
ALFRED: Signifying! Where are you at, man? Trying to find out, just find out. Have a right to know what you were doing there. You're off your beam, man.
PETER: Signifying—worse than stealing. Trying to find out about where I got the stuff. Son of a. . . .
ALFRED: Drop dead!

2

MR. L.: Come in, come in. Sit down.
ROGER: (*Walks over and sits down on the chair next to Mr. L.'s desk.*)
MR. L.: Now, let me see. You were told to come here by your teacher, Ms. Perry. No, that's not her name, it's Ms. Percy. Or is that it? Well, doesn't matter. Now, let me see. Oh, yes, you haven't been doing very well, have you? Clowning, hmmmm. Not studying . . . aha! And you have fallen asleep in class. Well, well. And you also have been annoying what's her name, Margie, drawing odd-looking pictures of her. So that's what it is.

3

MELANIE: (*Enters the nurse's office and sits down on the chair next to the nurse's desk.*)
MS. R.: Hello, Melanie. How may I help you?
MELANIE: (*Slowly moves her head from side to side.*)
MS. R.: (*Remains silent for a few seconds, meanwhile noticing that Melanie's eyebrows are drawn together across the bridge of her nose so that sharp lines cut into her forehead.*) Hard to begin?
MELANIE: (*Shakes her head up and down.*)
MS. R.: (*Smiles and places her hand on Melanie's hand, which is resting on her desk.*)
MELANIE: (*Begins to weep. Ms. R. gives her a tissue.*) I'm . . . I . . . I think . . . oh, no . . . can't. . . .
MS. R.: (*Remains silent but leans forward slightly.*)
MELANIE: I think I have syphilis.

4

SAM: (*Squirms in his seat, looks around, and then settles back with a vacant stare in Mary's direction.*) This just can't be. I feel so good, yet so bad—warm, cold. How long can it last? This magic of discovery . . . then distress of finding . . . I speak and no one hears. I feel and no one cares. When will there be peace? Peace of knowing what I want, where I'm going? Should we be together, alone, apart? Isolated in a crowd, crowded with thoughts when alone. I. . . .

5

ANGELO: I'm talking to you and you're not hearing.

FATHER: What do you mean? I hear better than you. You're just a know-it-all.

ANGELO: Just don't want to stay in college . . . a year or more. I want time to find out what I want. Don't have to worry about the draft; don't have to be drafted into education. Want to work, travel, be me. No *have-to*, just *want-to* every day.

FATHER: Know-it-all . . . this generation. All the advantages, takes . . . and takes. Don't appreciate the opportunities to get an education. In my day, my parents couldn't afford to send me to college. If only. . . .

ANGELO: You've got to listen. I'm telling you that life is more than college. Education is life. Life is education. College is not real.

FATHER: All the advantages, all they want to do is ruin their lives. This generation don't care about responsibilities.

ANGELO: Yes, I want responsibility—for myself. I want to find out what the world has to offer; I want to. . . .

This discussion of different forms of communication styles highlights the importance of interest in and awareness of the person at the end of the message. Helpful responses depend on skillful listening. Listening and responding are essential communication skills for the interchange of ideas. The effective interviewer must have these skills; the client should develop them.

Questioning and responding

The prime objective of the helping interview should be a movement toward trust and openness and away from anxiety and closed behavior. Getting into the center of the interview presupposes some degree of client/interviewer interdependence and mutual searching. To spur the client toward

greater self-realization the interviewer must be able to watch for nonverbal cues and to listen for verbal cues in addition to being aware of the sounds of silence.

The process of growth for both the client and the interviewer is further enriched by the interviewer's ability to formulate questions. Several considerations enter into the skillful use of questions.

The "when" and "how"

Often it seems that the interviewer thinks his interview work consists of bombarding the client with questions. The interviewer is in error if he relies too much on questioning; it is only one of many tools. The questioning tool becomes a weapon when the client is overwhelmed by the type, the frequency, or the procedure of questioning. The client is then likely to hide in his shelter of silence.

When questions are asked, they must be worded to correspond to the client's level of information. The client should understand what is being asked of him. The wide-awake interviewer knows that the question itself may be a subtle form of suggestion, permission, or challenge because of the way it is worded, by the tone of voice used, or by the particular words emphasized by the interviewer. For instance, the interviewer may be encouraging the client to do something when he says, "Did you *ever* tell your teacher what you *really* think of him?" On the other hand, the interviewer may challenge the client to consider—"Do you think another person would react as you did?" "Suppose you can't make it, what then?" "What if she doesn't care for you; what would you do?"

On the positive side, questions may be instruments to open the channels of communication because they energize thought and expression of feeling, redirect the client into some new considerations, or link the client's comments together. For example, questions may:

ENERGIZE THOUGHT: "What makes you think so?" "What is happening in your head right now?" "How do you explain his behavior?"

EXPRESS FEELINGS: "What makes you feel this way?" "What is happening between us right now?"

REDIRECT IDEAS: "I wonder what makes

you so concerned about how Kristen feels? Is it possible that you are thinking about something else, too?" "Is it possible that you talk about Paul to avoid talking about yourself?"

LINK COMMENTS: Client says, "My mother constantly calls me dumb. I guess maybe I don't work hard enough. But the work is too hard for me." Interviewer answers, "I hear something in what you are saying. I wonder, are you agreeing with your mother that you are dumb?"

On the negative side, questions can be an intellectualized way to cover up emotions, an attempt to pick at the client, a syrupy approach, or woolgathering in order to avoid threat or self-exposure. For example, questions are unproductive when they end up in:

PICKING AT THE CLIENT: "Now, really, do you think you should have done that?" "Who gave you the authority to go ahead with that plan?" "Come now, that wasn't right, was it?" "What *is* the matter with you?" "Why do you always say that?" "Why don't you ever do what I say?"

HEAD-TRIP: "I have been observing Susannah very carefully. Have you noticed her sybaritic (self-indulgent) behavior? How would you explain her constant requests for affection and food?" "Do you have any comments to make?" "I am aware of the disturbing symptoms your husband exhibits. Let's look at this calmly and come up with some reasonable process by means of which you may tolerate his peculiarities."

SYRUPY: "I am just here to help you find happiness. What I feel or think just doesn't count. What would you like me to do to help you?" "Yes, indeed, that's right. What would you like me to tell you [or do for you] next?" "Isn't that nice. I really feel you are doing so well you don't need me any more. Is there anything you want to talk to me about now?" "It doesn't matter to me, anything you say. What would *you* like to do this evening?"

WOOLGATHERING: "Hmmmm, you were just saying you find this chat unsatisfactory. Well, the goal we have today leads us further into discussion of your relationship. What about your relationship?" "You said your schedule is the same as Ken's. Did you discuss the book with your professor?" "So . . . you have to share a room with your younger brother. Do you have color television in your home?"

It is obvious by now that questions can go in many different directions. They can bring a client and interviewer together on the same wavelength or result in a confusion of sound.

When should questions be asked? Questions should be asked sparingly when they serve the following purposes:

INTERVIEWER OR CLIENT CLARIFICATION: "I'm sorry I didn't understand. Would you repeat that?" "I'm spouting forth. What do you think about what I've just said?" "I wonder what you felt when Delores refused to go to your senior prom?" "Would you give me an example of what you meant by . . . ?"

FOR EXPLORATION: "You mentioned that your father always made you feel unimportant. What do you mean?" "What further thoughts do you have about this . . . ?" "What are you feeling as we talk about this?"

FOR ADDITIONAL INFORMATION: "I think I got your message about Bud's reaction to his paralysis. When did Bud have his last operation?" "What are the required subjects in the major to which you have transferred?"

TO HEAR BETTER: "Sorry, I didn't hear that. What did you say?" "Did you say that . . . ?"

ENCOURAGE CLIENT TO TALK: "You said something about Barbara a little while ago. Are you interested in talking about her some more now?" "Your comments about Sid are very interesting. Let's get down to some gut-level feelings. How do you feel about Sid?"

How should questions be asked? Questions are asked as an invitation for the client to talk. They should be clearly stated in words that the client can understand. They should be as neutral as possible so that the words themselves have little influence on the nature of the answer.

Questioning and coping ability

Questioning and answering are processes in which individuals have participated all of their lives. The questioning age begins in early childhood. Children use questions as a means of attracting and holding adult attention and also to gain information. At first most of the children's questions begin with why. "Why" questions reach their peak in the second or third grade. Gradually the child adds "what" and "how" questions, and as he grows older, his questions become more specific. *Developmentally*, therefore, the individual employs questions to learn how to cope with the world.

The client's questions reveal a great deal about the way he views himself, events, and the people around him. Disordered individuals often reveal their *confusion* in the fuzziness of the questions they direct either to themselves or to the interviewer. Answers may be relaxing for some clients, but for the disturbed client, the answers just lead to more complicated questions. These clients frame their questions in such a way that they are unanswerable. They ask, "Why was I born?" "Why does everybody hate me?" "What is the meaning of life?" No matter how these questions are answered, they do not bring satisfaction to the client, for the client is not ready to accept solutions. He always finds some aspect still unanswered. If the interviewer tries to grapple with a question such as "Why was I born?" he prompts the client to ask many other questions, as exemplified by Mr. G., the interviewer, in the following situation.

MR. G.: That's a difficult question to answer. Each person finds his own reason for life.
STELLA: Yes, but how does one find a reason?
MR. G.: I can only tell you how I would find my reason. I would plan some goals and work toward them.
STELLA: Goals . . . but that's where I'm stumped. Who can tell about goals?

On and on this interview could go with no progress unless the interviewer confronts the client with the clear and frank statement that there is *no answer* to the client's question. In fact, the path to problem-solving might lead to a rewording of the question.

MR. G.: "Born?" There is no satisfying answer to that. But I do hear you saying, "How do I find my way? What are my goals?"
STELLA: Mmmmmmm. I think so. How do I know there is a reason for me to *be?*
MR. G.: Another hard question. It might help if you tell me some of the feelings you have when you ask that question.

Notice the different direction the revised interview is following.

Sometimes the client's question demonstrates his distress by *"ventriloquizing"* (Johnson, 1946); in other words, the client speaks with the voice of another. The client asks, "My husband wants to know when Jimmy is going to get over this identity thing you were talking about" or "Lots of people have been asking when you are going to make up your mind about Shirley."

Another coping device to avoid the reality of the moment occurs when the client asks historical questions. The client wonders, "Hey, Doc, do you think it was because of what happened when I was 16?" "When I was little, everyone thought I was such a sweet, obedient child." "If only my father had not hit me on the head when I was younger, I wouldn't have these headaches now. That's important, isn't it?"

Questioning serves many purposes. The interviewer's questions may lead to the discovery of solutions and to the linkage of events so that their meaning is clarified for both the client and the interviewer. The client's questions convey some understanding of his coping ability. Fuzziness in questions or nonanswerable questions are indicative of the client's level of development and his degree of distress.

Consideration of questioning moves to a new phase of inquiry, that is, the question of "why." The word "why" suggests surprise, probing, perhaps even prying, and also carries with it some negative effects.

The question of "why"

" 'Why' is a crooked letter." It is not only crooked, it is a hook that often pulls the listener into the fury of an emotional storm.

For the very young child, "why" is a searching word, an effort to understand. Soon the child finds out that when his mother says "Why did you do it?" she means that he was wrong and bad. Thus the "why" becomes corrupted. It loses its reasonable search for information and takes on a critical meaning.

"Why" frequently implies scolding, fault-finding, impatience, depreciation, and dissatisfaction as well as other harmful responses. When the student asks the instructor, "Why don't you lecture us on the principles of interviewing so we'll be able to apply them?" is the student interested in reasons or is he really saying "You have failed as an instructor"?

Thus "why," which in the origin of words asked for information, now is more likely to pose a threat to the client. "Why" prods and pushes and too often suggests to the client that he is to blame for something. When the client feels the disapproval of the interviewer's "why," he withdraws to escape the situation. The client defends himself instead of extending himself so that he may grow in further understanding. Exercise 6-22 has a list of "why" questions that indicate how these questions suggest disapproval.

Exercise 6-22
THE QUESTION OF "WHY"

Change the following list of "why" questions into questions beginning with "what" or "how," or better still, change them into statements rather than questions. Use the following example as a model for this exercise.

EXAMPLE: "Why did you say that?" Change to: "Would you explain what you meant?" "I hear you saying that you think I have been undermining your efforts. Tell me what I did to make you feel this way."

1. Why do you think your parents are not helping you?
2. Why do you come in here looking that way?
3. Why do you keep touching your hair while you are speaking?
4. Why do you act this way today?
5. Why don't you like this course?
6. Why don't you keep away from people like that?
7. Why are you always in such a hurry?
8. Why are you so disorganized?
9. Why are you silent now?
10. Why do you argue with me so much?

Sometimes the question of "why" suggests the "tyranny of the should" (Horney, 1954). The "should" interviewer becomes irritated with the rebellious client who does not accept the interviewer's directive remarks. This same interviewer is elated with the client who blindly follows the scheme of behavior presented by the interviewer. There are many subtle ways that "should" interviewers jab at the client. The interviewer asks, "Shouldn't you consider what others think of your behavior?" Don't you think you should consult your wife about that decision?" "Why do you think you should do that?"

Varieties of questioning

Closed/open questioning. "Do you want to learn about your problems?" Ob-

viously this question can be answered with one of two answers, "Yes" or "No," or perhaps a third answer, "Not sure." This is therefore a closed question. The client must select from a series of possible answers what best fits his wishes or feelings. Although the client has two options, "yes" or "no," these lead to a dead end. The question restricts the answer and is so narrowly focused that the client is not afforded leeway for self-discovery.

Rewording the question makes a difference: "Let's look at this mix-up together. How do you feel I might help you find out more about your problem?" The revised question allows the client fuller scope. Open-ended questions invite the client to voice his views and his feelings. In this way the contact between client and interviewer is widened and deepened since the client is included in the undertaking. Note the differences among the following:

"When did you first notice you were getting depressed?" "Tell me about your feelings . . . about your depression."

"Do you think your grades will be higher or lower this semester?" "How do you feel you are going to do this semester?"

"Do you feel your family is helping or hindering you with your decision?" "Tell me about your family. How do they enter into your decision?"

Closed-ended questions readily display their factual, more focused purpose, while open-ended questions leave more space for the client to structure his own answer as he sees fit. The effective interviewer uses both kinds of questions, leaning a little more heavily on the open-ended questions. Both kinds of questions have their advantages and disadvantages, depending on the objectives of the interview. The more anxious client may feel more comfortable beginning with a more defined (closed-ended) question and answer that require less effort and less involvement. As he begins to feel more accepted and worthy, he will be able to handle the looser, open-ended question. On the other hand, the reluctant client may be less threatened with the open question that provides freedom to move as he pleases.

The skilled interviewer knows how and when to use both open and closed questions. He is more likely to start the interview with an open-ended question such as, "How have things been going with your roommate this past week?" The interviewer is careful to interject some structure in the form of closed questions with the rambling client who appears to be avoiding the issues. Yet, since listening to the client is the essential of the interview, the interviewer abstains from the comfort of frequent closed questions. Closed questions force the interviewer to concentrate on what to ask next rather than taking heed of what's happening with the client.

Closed-ended questions and open-ended questions are related to two other characteristics of questions—leading/responding questions.

Leading / responding questioning. Closed questions are constructed to lead. Open questions follow the client's lead and thus are responding to the client. In the first instance the interviewer functions as a guide for the client. The interviewer controls what happens during the interview, and because he maintains himself in the center of things, he often suggests the client's answer. Frequently leading questions become a cross-examination in which the interviewer implies that he already knows the answer. The interviewer says, "You loved your father, didn't you?" or "You're 34 years old now, aren't you?"

If the interviewer is asked his reasons for wording the questions as he does, he may justify his approach with the assurance that he is well informed about the client's level of understanding. This may actually be a cover-up for the fact that the interviewer enjoys directing the client. Another reason the interviewer may offer is that he thinks the client is wandering from the topic too much or is getting bored. Once again, this may conceal the interviewer's discomfort or his inability to get interested in the client's concerns.

At other times the leading question helps the client who is confused and vague to pin down a particular idea or feeling. In such a case it is similar in value to the closed-ended question, that is, to be used discreetly in times of need.

Contrary to leading, responding to the client impels the interviewer to speak in the client's terms and to follow the ideas and feelings the client communicates. Rather than the interviewer leading the questioning, responding demands that the questions place the client in the center. The interviewer whose philosophy is that the client has solutions sets about helping the client work out his problem.

Instead of saying to the client, "Tell me whether you are for or against amnesty for deserters from the Vietnam War," the interviewer asks, "I'd like to know more about your thinking about amnesty." He encourages the client by his one-word comments or his well-timed pauses to talk a little more about what he (the client) has just briefly mentioned. The interviewer nods his head and shows by other body movements that he is interested in what the client is saying. He says, "I see. Would you explain a little more what you are thinking about?" "Ummmmm-hmmmmm." "Tell me more about it." "Not sure I understand. Did you say . . . ?" "You said that you didn't like the way John answered. Suppose Suzie said the same thing?"

In essence, responding questions follow the client's leads; leading questions prompt the client to follow the interviewer's cues. Leading questions have much in common with the next category of questioning—answer/agree questions.

Answer/agree questioning. The answer/agree question is another form of the leading question. The interviewer asks, "You didn't mean to do that, did you?" or "You were upset and lost control, didn't you?" The question includes the answer and the interviewer expects the client to agree with him. The client may comply to maintain the goodwill and warmth of the interviewer and thus avoid his displeasure and rejection.

A question that demands that the client go along with the interviewer is also a closed question. The client is boxed in when the interviewer says, "No one would rip-off unless he knew why, would he?" "It must be perfectly clear to you why your father lashed out at you as he did . . . after you did that. Isn't it?" "Surely you can understand that you should keep away from your old crowd. They'll drag you down again, won't they?" "Just because you're angry with me right now doesn't mean you have to leave right away. Do you?" "Come now, you know your mother loves you, don't you?" Often these questions require no more than a "yes" or "no" answer from the client, and they are more likely to be stated in an impersonal way.

If these questions are opened up, they permit the client to give *his* answer, not the built-in answer of the interviewer. "What are your thoughts about what makes you rip-off?" "Let's talk about your father. After you . . . , what do you think he might have done?" "Have you considered what would happen if you returned to your old crowd?" "I feel that you are angry with me. It would help us both if you stayed and we worked this out together." "I see how you feel about your mother. Do you think perhaps that's the way she shows her love?"

Answer/agree questions have features similar to closed-ended questions, to leading questions, and also to direct questions, which are discussed next.

Direct/indirect questioning. There is a place for a direct question in the interview when the interviewer requires some information immediately or when the client is floundering or anxious and needs some additional interviewer support. On the other hand, the indirect question is preferable since it puts the client in charge of his answer. Direct inquiries usually have a question mark at the end of the statement. Indirect inquiries explore without a question mark at the end. The following remarks contrast direct with indirect queries.

"What do you think about . . . ?" "You must have many thoughts about. . . ."

"How do you feel about . . . ?" "I'd really like to hear your feelings about. . . ."

"It must be burdensome to attend college during the day and to work at night. How do you do it?" "That takes some doing, working at night, college during the day . . . tell me about how you manage it."

"Does your stuttering bother you very much today?" "I'm really interested in what's happening in your head today."

The concept underlying the unmarked question is a cooperative searching by interviewer and client. The direct question ("What procedures have you considered for changing her opinion of you?") may open more response doors for the client than the closed question ("Do you have any procedure for changing her opinion of you?")

However, when the interviewer learns how to converse effortlessly without the hook, the question mark, he gets closer to a meaningful sharing with the client. He is also less likely to become involved in questions sprinkled with multiple ideas.

Exercise 6-23 differentiates between direct questioning and indirect questioning. Direct questioning, when compared to indirect questioning, is more likely to be leading and controlling.

Exercise 6-23
NONCONTROLLING QUESTIONS

In order to distinguish between leading (controlling) and nonleading (noncontrolling) questions, select a subject about which you wish to get some information. Decide on the information you want to get (objectives). Write five leading (controlling) questions that would get the information you want. Then ask someone to answer these questions. Write down his answers to these questions.

Use the same subject and objectives as before. Write five questions that are not leading (noncontrolling). Ask the same person the revised questions. Write the answers to the questions in the answerer's words.

What were the characteristics of your questions that made them leading (controlling)? What were the characteristics of your questions that made them nonleading (noncontrolling)? Discuss with your client how he felt about the two sets of questions.

Multiple-idea/single-idea questioning. Questions may take in a large territory of ideas but should have only one reference point. Thus the interviewer may inquire "How do you feel about your field work (practicum) experience?" rather than "Do you like your field work placement and your supervisor?"

The difficulty with the multiple-idea question resides in the client's inability to decide which question he should answer first. Sometimes the client wants to answer only one part of the question and would be uncomfortable in replying to another aspect. Consequently, he may not respond or go off on a tangent that has nothing to do with the original question.

Multiple questions such as the following get too complicated for a reply: "Do you see interview behavior as an action-reaction, interaction, or a transaction?" "Do you believe that, if you do no harm to a client, you have done some good for him, or is there some aspect of interview withholding involved here?"

Even questions that present one choice out of two present some hitch. It would be much simpler to answer two questions—"Do you like college this semester?" and "Do you like your professors this semes-

ter?"—than to reply to "Do you like college and your professors this semester?"

A seemingly easy question such as "When did you first notice you were getting tense, before or after you came here?" becomes cumbersome because it suggests that the client must narrow himself to one of two choices and also must reply in terms of the time of tenseness, not the meaning or extensiveness of it for him. Besides, the client

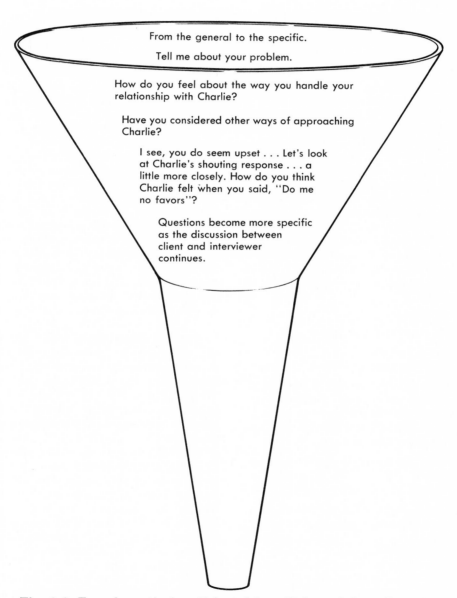

From the general to the specific.

Tell me about your problem.

How do you feel about the way you handle your relationship with Charlie?

Have you considered other ways of approaching Charlie?

I see, you do seem upset . . . Let's look at Charlie's shouting response . . . a little more closely. How do you think Charlie felt when you said, "Do me no favors"?

Questions become more specific as the discussion between client and interviewer continues.

Fig. 6-2. Funnel questioning. (Adapted from Kahn and Cannell, 1957.)

may answer "I don't know" rather than explore further because he is unable to pinpoint the originating moment in terms of "before" or "after."

The intent of this discussion of questioning is to make the question the servant rather than the master of the client. Therefore the interviewer is charged with the responsibility to develop a more comfortable feeling about his interrelationship with the client using open, responding, answerable, indirect remarks as often as possible.

Funnel/inverted funnel questioning. Before completing this examination of questioning, one additional procedure is discussed. This procedure is concerned with the process of questioning from generalities to specifics. Kahn and Cannell (1957) write of the difference between the "funnel" sequence and the "inverted funnel" sequence of ques-

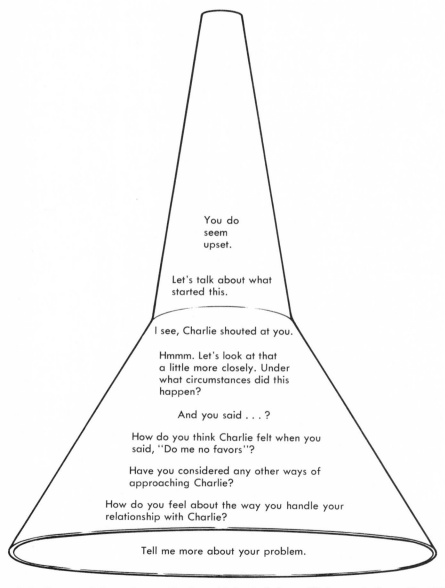

You do
seem
upset.

Let's talk about what
started this.

I see, Charlie shouted at you.

Hmmm. Let's look at that
a little more closely. Under
what circumstances did this
happen?

And you said . . . ?

How do you think Charlie felt when you
said, "Do me no favors"?

Have you considered any other ways of
approaching Charlie?

How do you feel about the way you handle your
relationship with Charlie?

Tell me more about your problem.

Fig. 6-3. Inverted funnel questioning. (Adapted from Kahn and Cannell, 1957.)

tioning. The funnel sequence indicates that the interviewer asks questions from general areas to more specific points. In other words, beginning questions are less restricted, and as the interview goes on, questions become narrower and more informationally related to interview goals (Fig. 6-2).

Several characteristics emerge from the questions listed in Fig. 6-2. The first question listed is an open, indirect remark that aims to bring to light the client's point of view. ("Tell me about your problem.") The client has the chance to present those ideas or feelings that he considers most important. As the conversation progresses, the interviewer grasps the client's feelings and ideas and narrows his questions to the more specific issues that seem to be involved. ("How do you think Charlie felt when you said, 'Do me no favors'?") Gradually more specific issues and perhaps some closed questions may be used in order to help the client perceive his own behavior in relation to what is happening around him.

The opposite procedure is the inverted funnel sequence. In this approach, questioning urges the client to think through subareas of his primary concern. The questions move from the specific to the general (Fig. 6-3). The latter sequence, from specific to general, helps the client gather points that may be influential in his problem. Then he is helped to tie them all together to reach some solution. The particular sequence that is best depends on the client's and the interviewer's learning styles (p. 7) as well as the objectives of the interview. The more experiential the interview, the more likely that the preferred procedure will be the funnel approach. The more data-gathering the interview, the more likely it will proceed from specifics to generalities, the inverted funnel approach. In Exercise 6-24 the dependence of the client's response on the interviewer's questioning and responding skills becomes apparent.

Exercise 6-24
QUESTIONING AND RESPONDING

One trainee is selected or volunteers to go out of the room. Before he or she leaves, the following situation is described. The role-player has worked in an office for the past 4 years. His work record is efficient and he thinks that he is next in line for the job of office manager since the former manager has resigned. He plans to speak to the personnel office about the possibility of getting the job.

After this explanation, the role-player goes out of the room to plan how he will handle the request for the job from the personnel officer. With the role-player out of the room, a personnel officer is selected and the group with the personnel officer plan how they will approach the situation when the role-player asks for the office manager's job. Their plan should include one of the following problems that would prove unfavorable for the role-player to be promoted to the position of office manager.

1

Role-player does not get along with the other workers. In fact, some of the workers have come to the personnel officer to complain about role-player's brusqueness.

2

Role-player has a peculiar body odor that some individuals find offensive.

3

Role-player has been absent a great deal recently. Although he functions well when he is working and makes up for his absences, the personnel officer is wondering what has caused the frequent absences.

• • •

The group selects one of these problems and with the personnel officer plans some questions and areas to examine. Then the role-player is called back into the room.

The personnel officer conducts a 5-minute interview. While the personnel officer conducts the interview, the other trainees keep a record, such as that given on p. 140.

After the interview the group discusses for 10 minutes the predominant types of questions used and the favorable aspects of the procedure the interviewer followed. Then suggestions are made for improvement.

The interview is repeated, including the tallying of questions and the 10-minute discussion, with the additional consideration of improvements of the second interview over the first one.

ITEMS	TALLY NUMBER OF EACH
Number of open questions	
Number of closed questions	
Number of leading questions	
Number of responding questions	
Number of answer/agree questions	
Number of direct questions	
Number of indirect questions	
Number of multiple-idea questions	
Number of single-idea questions	
Approach primarily funnel sequence (check this or item below)	
Approach primarily inverted funnel sequence	

ENDING OR TERMINATING PHASE
The single interview

Often starting and ending an interview are more difficult for the beginning interviewer than developing the interview once it is started. Time is an ally in both of these instances. The interviewer may begin with, "We have 50 minutes to rap today. Where do you want to start?" The interviewer may end with, "Ten minutes more. Hmmmm. Let's look at what's been happening."

Time boundaries

A natural boundary line for each interview is created by the element of time. If the interviewer has initially stated that there are a certain number of minutes to the interview, this serves as the time boundary. Sometimes this time boundary may be imposed by the number of minutes allotted in a school period or by the work setting.

In spite of this time boundary, the interviewer is often concerned about how to ease the client into the home stretch. This may be of particular importance if a series of interviews has occurred and the question of closure (completing the pattern of the interview) arises.

Timing devices

The beginning interviewer may watch his clock more often in order to determine when 10 or 15 minutes remain for the interview. As the interviewer becomes more proficient,

the interviewer develops a built-in timing device based on the cues from the client and automatically becomes aware of when the time boundary has been reached. The client also becomes alerted to time signals. This framework of time may be a comfortable boundary for some clients; for others it may be a blunder. Cultural and other unique characteristics of the client influence the meaning of time. The empathic interviewer knows the relative meaning of time for the client and pursues the most effective procedures possible within the framework that may be imposed on him. The time boundary may also impress the client with more responsibility to perform his interview work.

Assuming that time boundaries are imposed by outside requirements and cannot be changed in accordance with the client's needs, the 10-minute wrap-up period is a favorable procedure. No new material should be begun during this period. Instead, the client should be assured that his new thoughts may be discussed with the interviewer at the next interview or the client may be referred to someone else if the material seems important and only one interview is permitted.

The introduction of new material when another client may be awaiting the interviewer's attention might result in a hasty and careless examination. Both client and interviewer are apt to become irritated.

Even if the information is what the interviewer has been hoping the client would relate, rushing through the discussion of this disclosure is unfair and unsatisfying to the client.

Concluding remarks

Closing remarks may come from either the client or from the interviewer. The client indicates that he is ready to leave by comments that suggest he has shaped his thoughts into some ideas for action. The client says, "That's it. I know what I want to do now. I will . . ." or "This chat has given me some things I want to think about before I decide what to do. See you next time, Mr. H."

The interviewer's comments seek to give the client the opportunity to perform the finishing touches, to plan some actions the client should accomplish in between the interview sessions, or to plan for a subsequent session. Some of the concluding remarks of the interviewer might be, "We have about 12 minutes left in this interview; are there any questions you would like to ask?" "In the remainder of the time we have left today, is there anything in particular you think we have to pull together?" "In the time we have left, let's see if we can pull together the main concerns we have talked about." "Our session is almost over; would you like to think about what you might be doing after this interview?" "Our session is almost over; let's think ahead to what we might want to talk about at our next meeting."

Nonverbal endings

Nonverbal endings often speak louder than concluding remarks. When the client begins to shift around in his seat, looks at his watch, or even more obviously, reaches for his car keys, these signals come through clearly if the interviewer hangs in there with the client.

The interviewer may also begin the wind-up process with movement cues. He may straighten the papers on his desk, arise from his seat, or look at his watch. Depending on the age and developmental status of the client, the interviewer may hold out his hand to shake the client's hand or touch him on the shoulder as a form of departure. Some form of courtesy in leave-taking is a wise

interpersonal policy. Such courtesy as well as the verbal preparation for termination serves to assure the client that he is not being pushed out and thus rejected.

The difficulty of parting may be lessened if the interviewer is not fearful of separation from the client or anxious about losing the client's confidence. The important factor is that both the interviewer and the client must be prepared for terminating the interview.

Interview series

Dealing with a series of interviews presents some similar problems of closing each individual interview and some additional problems.

Number of interviews

In some instances an agency or individual interviewer sets forth in the beginning interview, the number of interviews that will be included in the series. Thus the interviewer may latch on to the already agreed on final interview, and during the next-to-the-last interview the interviewer prepares the client for the termination. The interviewer reminds the client of the forthcoming ending interview.

When a series of interviews is indicated, the termination of the series does not stamp the client's problems as having been completely solved. Instead, the client still has to work on positive growth as well as continue to achieve appropriate coping behavior. When the client is ready for this self-sustaining growth, the empathic interviewer perceives this and removes himself before he becomes a heavy crutch.

"Hello-goodbye"

Occasionally the client leaves too soon. Brammer and Shostrom (1968) speak of the "hello-goodbye" or "flight-into-health" phenomenon that a client may experience after only one or two interviews. The client feels relieved, relaxed, "so good." He believes his problems have been solved because for a brief period of time his conflicts and distress have decreased in annoyance value. However, from the client's behavior the interviewer realizes that the client is leaving by a revolving door. The client's return to the same or to another interviewer is bound to happen.

Sometimes nothing can be done to encourage the client to remain in the interview situation for a few more sessions. When the client decides to check out, his determination has to be respected. The only further help the interviewer can offer is to assure the client that he may return if he wants to do so.

Closing devices

When the client becomes more certain in his discussions about himself, he will begin to question the interviewer about ending the interview series. This client-determined termination is, of course, the simplest to accomplish. The interviewer need only agree: "It seems that you can operate on your own now. You've shown you can solve your problems. You're aware of when and in what way certain events trouble you. How about one or two more times together to tie things up?"

For the client who is lonely or dependent, depriving him of the interview relationship might be too stressful. Therefore the interviewer begins a gradual weaning process. Interviews are arranged at longer and longer intervals rather than cutting them off suddenly. If the interviews have been conducted once every week, then the spacing might be changed to every other week. Later on, depending on the anxiety level of the client, the interval between interviews may be lengthened.

Another effective device that keeps communication channels functioning even between interviews is to ask the client to write a summary letter. After a certain period of time, the client reviews in writing what has been brought to light since the last interview and what growth the client considers still necessary.

Open-door policy

The abruptness of a feeling of dismissal may be avoided also if the interviewer tells the client to "Drop in sometime when you're near the office" or "Let me know how things are going." Thus the interviewer keeps the door open so that the client knows that he may contact the interviewer in times of crisis or even for a friendly chat.

Caution is necessary so that this open-door invitation for further contact does not become a pass to further dependency. The empathic interviewer understands the client and adjusts his offer accordingly.

Referral

At times referral to someone else for further interview sessions may be desirable. If only a set number of interviews are permitted with one interviewer or if the interviewer feels that because of certain problems in the interviewing relationship the client and interviewer have gone as far as they can together, then referral may be the wisest course to follow. In order to avoid having the client feel as if failure has resulted, that he is being rejected, or that he is getting the "run-around," the interviewer must discuss the reasons for referral in such a fashion that the client does not feel depressed or frightened. Suggesting some possible other interviewers or perhaps someone who specializes in some aspect of the client's further developmental needs builds a bridge for the client.

Summary of Part Three

Varieties of interview orientations were discussed in Chapter 5. Although each variety of interview envelopes some distinguishing characteristics, there is also an inclusive wrapper extending over all forms of interview that covers the interviewer's ethical responsibility and his interest in the client as well as the form of communication process.

The information-oriented interview is designated for obtaining and/or sharing information. This interview is usually more structured in procedure and goals than the experiential interview. Often what is required is an answer to predetermined questions by the interviewer, by the client, or by both of them.

The behavior-modification interview focuses on positive reinforcement of appropriate behavior and the nonreinforcement or negative reinforcement of inappropriate behavior. The interviewer must shape his own behavior so that he will systematically reinforce the established behavioral goals. The steps in a plan for behavior modification require behavioral analysis, a reinforcement menu, and the terminal behavior that is the goal of the behavior modification.

The experiential interview incorporates information and behavior change but in an atmosphere in which human relations are the crucial elements. The interviewer himself becomes an important part of the process that (1) seeks to guide and facilitate (make easier) the client's exploration of his behavior, concerns, and problems in the "here and now" of the interview situation; (2) strives to establish a miniature (microunit) social situation in which the client may try out interpersonal relations as well as some solutions to problems; (3) sets up an acceptant atmosphere of trust in which the client feels safe to try out some changes in behavior; and (4) provides another individual who listens, participates, and becomes an involved helper on whom the client may depend and then grow away from as the client's self-confidence increases.

In Chapter 6 the three overlapping phases of the interview structure have been distinguished. The opening phase of the interview sets the stage for a productive and harmonious relationship. At the beginning both client and interviewer clarify their roles and relationships. When both client and interviewer arrive at an understanding of the "interview work" they must accomplish together with the purposes and expectations of the interview, they commence to ease their way into exploration of what has become mutual concerns. Several possible pitfalls may disturb the helping relationship. These pitfalls may take the form of interviewer manipulation of the direction and content of the interview. Difficulties also may arise from either the personality oddities of the interviewer or his lack of observational abilities and inadequate communication skills. The management of the initial greeting, setting limits, and handling feelings depends on the empathic understanding of the interviewer. The genuineness of the interviewer's commitment to the client and the nonpossessive warmth he exhibits during the interview are additional building blocks for the personal growth of the client.

The discussion of the developing or exploring phase concentrated on techniques for spurring the client on. Basic to the smooth interchange on which the effective interview is built is the attending behavior of the empathic listener. If the interviewer listens to the client's words, to the client's body language, and to the sounds of silence, he can respond more capably to the client. The "when" and "how" of questioning insists on the interviewer's alertness to client cues along with regard for the characteristics of questions. The client must be at the center of the interview at all times, even when the interviewer is speaking. This means that the flow of conversation as expressed in the client/interviewer dialogue follows the client's pace and the client's lead.

Terminating the single interview or a series of interviews requires client preparation and the avoidance of a feeling of eviction. The client and the interviewer make themselves ready for the separation. By means of mutual feedback, the client and the interviewer make known new understandings and how the client has translated these notions into action. If this is the first of more than one interview, then the interviewer reinforces the client's search for some goals to work on between interview sessions.

Closing devices may include an "open-door" policy. The open-door invitation offers a helping hand that reaches beyond the ending interview. However, this invitation must be used with caution so that the client is not prodded into a dependency relationship. In addition, since the client continues to develop more self-fulfilling, appropriate behaviors, he should not be encouraged to use the open door as an excuse to run to the interviewer for assistance rather than working out his own problems.

Some considerations that might ease the closing of either a single interview or a series of interviews are time boundaries, verbal and nonverbal endings, recapping accomplishments, decreasing the frequency of visits, and referral.

PART FOUR

FROM INTERVIEWING TO COUNSELING

THE CLIENT IS ON HIS WAY TO SOMEWHERE

CHAPTER SEVEN
EXPRESSIVE SKILLS OF COMMUNICATION

Confrontation. (Photo by Irwin Nash; from Wagner, N. N., and Haug, M. J., editors: Chicanos: social and psychological perspectives, St. Louis, 1971, The C. V. Mosby Co.)

The status of the helping person is not as important as the helper's belief that people have the capacity to grow. Therefore the interviewer, the counselor, and the psychotherapist may differ in characteristics such as their level of training and extensiveness of experience, in the functions they perform, and in the settings in which they work, but they all must develop the same basic skills of observation, listening, and responding. In addition to these basic skills are the more specific communication skills that the counselor uses to clarify, to convey the meaning or feeling, and to indicate that he has received the client's message accurately.

These expressive skills are paraphrasing and summarization, reflection, confrontation, and interpretation.

PARAPHRASING AND SUMMARIZATION

The interviewer has to learn what the client is conveying. This does not imply that the client's word meanings are the same as the counselor's but that the counselor is able to listen so well that he understands the client's frame of reference. In other words, he is on the client's wavelength. The alert counselor makes every effort to understand and to share his understanding with the client.

Paraphrasing and clarification

Paraphrasing is a form of translation. The message receiver (the listener) rewords what he has just heard to pull together the chief points. The primary purpose of paraphrasing is to make clear what has been said by restating what the message sender (the speaker) has said. Repeating, highlighting, and coordinating are interrelated characteristics of paraphrasing.

Repeating

In some instances, paraphrasing merely repeats.

MANNY: I have a galloping case of depression. Emptiness just grows. The hollow yearning for something runs through me more and more frequently. There are no more daily dividends of joy, not even my quiet feeling. I feel remote, even here with you. Confused, just confused. A prisoner of my depression. *(Stops talking suddenly and leans forward to look at the counselor, Ms. R.)*

MS. R.: Depression? (Or "Prisoner of depression?")

Paraphrasing by repeating is the least complicated method used to focus the client's attention on certain aspects of his communication. By repeating one or several words the counselor shows that he is listening. He also steers the client to continue a particular direction of discussion.

A flow of repetition, however, results in a dull and probably less effective session. Clients who are exposed to too frequent repetition sometimes ask "Don't you hear me?" or "Something wrong with the words I use?" A greater variety of responses requires deliberate listening and keener "headwork" on the part of the counselor in order to catch the feeling and meaning of the client's remarks.

Highlighting

Another form of paraphrasing requires that the counselor concentrate on the client's train of thought. In highlighting, the counselor emphasizes the main points in several of the client's last remarks. These counselor comments may not be in the sequence in which the client has just expressed them but in accordance with a theme the counselor has begun to notice. Ms. R. might respond to Manny by saying:

MS. R.: Hmmmmm. A galloping case of depression, confined by depression. No more joy, just isolation and emptiness.

By means of highlighting the counselor pulls together and repeats several of the client's words or phrases. The difference between shallow repeating and purposeful highlighting lies in detecting the recurrent theme of the ideas in the client's last statements. Highlighting serves to assure Manny that Ms. R. is trying to see events through his eyes. It also draws Manny's attention to the central concerns in his statements.

Coordinating

Contrary to repeating one or more words or to highlighting the main points of the client's statements, coordinating is the process of organizing and rewording the client's message. The counselor tackles the coordinating procedure cautiously so that the meaning of the client's statements are kept intact. Thus Ms. R. would coordinate Manny's outlook as follows:

MS. R.: Sounds like you're telling me that your depression is increasing rapidly and is becoming more annoying . . . and you are seeking something—something that will break up the emptiness with a comforting feeling.

Changing some of the words may have altered the poetic sound of Manny's statement, but it still carries the meaning of his comments. Ms. R. emphasizes Manny's increasingly annoying depression and his hope for a change in his feelings. Although she may not comment on the way in which Manny presents his thoughts, that is, his particular choice of words, she remembers and records these words. These words may have some bearing on Manny's distress and on the quality of his contact with social realities, even though at this counseling moment she is unable to determine the significance of his words.

Exercise 7-1
PARAPHRASING

MS. H.: Ms. M., I've heard about you from my friend, Zita Camshun. She says she belongs to the same women's liberation group as you do. Been reading books about women's liberation movement and am more and more angered by the marketing outlook of many, no, I mean most, men. "What's a woman got to give?" "Why should a woman take away a man's job?" That's what I hear them saying. So my consciousness is raised. Now I see how I have been exploited by the masculinist. Hell, I've got ideas. I've got skills. Not feminine ideas, not feminine skills— just ideas, just skills . . . no sexism with them. So where am I at? Beginning to be torn between husband, home, children, and me. Does it have to be this way?

Paraphrase Ms. H.'s statements by means of repeating, highlighting, and coordinating. Discuss your responses with others in your training group.

Summarization

Summarization overlaps with paraphrasing. Clarification is sought by feeding back to the client the substance of what he has been talking about. The primary difference

between paraphrasing and summarizing is in the amount of material covered. Paraphrasing concentrates on the client's immediate comment; summarization restates what the client has been saying during an entire counseling session or a series of counseling sessions. Another difference is that paraphrasing concentrates on the idea content rather than the feelings expressed while summarization pulls together ideas, feelings, or both.

Summarization of content

The counselor may summarize the themes and thus the meaning of the client's discussion. This summarization may occur at the beginning, at the end, or sometimes during the counseling session.

At the beginning of the session the counselor starts the conversation by recalling the main ideas of the previous interview or series of interviews. The client is urged to continue where the prior session stopped. Summarization at the beginning of a session is of particular value if the client is having some difficulty in starting the conversation. This review also aids the client in noting what he has expressed so far on a particular topic.

Either the client or the counselor may summarize. When the counselor seeks to encourage the client to summarize, he might say, "Do you recall what we talked about the last time we met? Try to recap it. It will help us go ahead from our ending point." When the counselor decides it would be best for him to summarize, he might initiate his summary with, "If I recall correctly, you were saying at our last session. . . ." The counselor prefaces his remarks with "If I recall correctly . . ." or "If I understood what you said . . ." to leave the way open for the client to correct any errors in the counselor's understanding.

Summarization of feeling

Accuracy in condensing the feelings the client expresses during one or more sessions depends on whether the counselor has been plugged into the client's wavelength. One difference between summarization and reflection of feelings is the time span and the range of feelings covered. Both reflection and summarization of feelings pursue clarification for the client and the counselor. In summarization, however, the feelings discussed involve an entire session or series of sessions and also include a wider variety of feelings.

Empathic understanding is necessary for this review of feelings. The counselor who is tuned in to the client notices the consistent patterns of both positive, negative, and/or ambivalent emotions (Chapter 6). When summarizing, the counselor must be wide awake and sharp-witted so that he correctly echoes the client's feelings.

Since errors in summarizing feelings are possible, the wise counselor adds to his summarizing remarks: "I think you have been telling me you feel. . . . Does it seem this way to you?" or "The feeling you expressed in the last few sessions made me think you. . . ."

REFLECTION

Recognition and reflection of feelings may help the client become aware of the degree to which his behavior is molded by his feelings. Reflection is a very difficult expressive response to accomplish and should not be attempted by the untrained individual. Accurate reflecting is dependent on developing the skills of attentive listening and empathic understanding.

Explanation of reflecting

Reflecting is often referred to as the "Rogerian mirror." The technique of reflecting feelings was first clarified in Carl Rogers' book *Counseling and Psychotherapy* (1942).

Definition

Reflection is more than echoing the feelings that the client has put into words. It is the act of uncovering and making known in fresh words the feelings that lie within the client's comments. Thus the counselor brings the client's feelings to the client's attention without adding or subtracting from what the client is saying.

TOM: All I keep thinking is "Man, you blew it. You got the dream but not the drive." I'll end up hangin' 'round the shopping center with all the other freaks.

MR. A.: I hear you saying that you're feeling pretty hopeless at this point.

Reflection and other expressive skills

Reflection borders on summarization of feelings and confrontation. Reflection of feelings is a step beyond summarization since in reflection the counselor feeds back those feelings the client has just labeled and also labels the intent of the feelings the client has described in his ideas. Summarization of feeling merely pulls together the feelings that the client himself has labeled during one or more sessions. The following is an example of reflection of feeling.

STAN (a Vietnam veteran): All I want now is to erase their faces as we shot them down. Now, I want to forget their twisting and turning, their death dance. To forget the shot of pleasure when I killed my first 15-year-old for democracy.

DR. R.: Yes, I know what you mean. I was there. The guilt gnaws at you.

If Dr. R. were summarizing Stan's feelings, he would have waited until Stan stated his theme of guilt over a longer period of conversation. Assuming that Stan has been speaking of feelings similar to those in his preceding statement, Dr. R. might summarize the feelings as follows:

DR. R.: I hear you saying that you want to *forget*, to forget the look on their faces as they died, the pleasure you felt when you killed your first 15-year-old—forget the memories.

A form of reflection that goes beyond reflection of feeling and approximates confrontation is reflection of experience. The reflection of the client's experience focuses on contradictions the client discloses between what he says and what he shows by his nonverbal cues. The client is faced with the discrepancy between his words and feelings. Confrontation delves further into the similarities, differences, and confusions that the client is and has been expressing in his ideas and feelings over one or more sessions.

Types of reflection

Reflection "puts feelings on the line" so that the client may become more aware of them as well as realize that it is okay to feel. The counselor works at sending the client the message, "I am here to share the burden of your feelings with you until you are prepared to shoulder them alone. I am with you, trying very hard to understand how you feel." In this way, reflection increases the client-counselor involvement.

Reflection of feelings. Recognition and reflection of feelings assist the client to inspect his feelings and to note how they relate to his behavior. The counselor's empathic understanding and skillful approach in reflecting these feelings are of paramount importance. Not only may possible distortion irritate the client but even suitable and accurate verbal reflection may result in an increase in the client's unpleasant feelings. When the counselor's reflection is acceptable to the client, it supports the client's further exploration of his feelings.

The client hears his feelings discussed twice. First he remarks directly and/or indirectly about his feelings. Then the counselor, who accepts and respects him, describes the client's feelings. This double exposure helps clear up any undercurrents of sadness, anger, frustration, and dejection suggested in the client's word content. Reflection of feeling provides opportunity for the client's self-understanding but is not the primary goal of counseling. The client must learn both to trust the expression of his feeling and to control his emotional behavior.

When the counselor reflects the client's feelings, he usually precedes his remark with "You feel," "You think," "You seem to feel," "You sound as if," or "I hear you say." The counselor, Mr. A., illustrates reflection of feeling in his comments to Bonnie.

BONNIE: How do you tell a big-shit, white, businessman churning it up in a motel that you're not for hire . . . working as a maid, a black maid, cleaning up his damned mess. Then he offers me—*(Begins to sob.)*
MR. A.: Angry, I hear your anger and I hear frustration. I do understand. Are you also asking how to handle a white man's prejudice?

From the moment that the counselor begins to reflect the client's feelings, he assures the client that he is not just a spectator in this interpersonal transaction. Often the counselor goes a step beyond the role of spectator, to that of participant and sharer of experiences.

Reflection of experience. In reflecting experience the counselor advances to more complex helping. He tells the client, "I am listening to you very carefully. I am looking at you intently. I not only sense your feelings but I hear and see that what you say and what your eyes, hands, body do just don't fit together." By means of this thought process the counselor becomes more than an echo or a mirror; he now shares his own experiences with the client.

Other terms that may be used for reflection of experience might be "congruence" or "genuineness." All three refer to the consistency between what the client says and what he implies by his facial expression, his gestures, and his tone of voice. Inconsistency between the client's or the counselor's verbal behavior and his nonverbal signals often indicates that the individual is not prepared to be open about some aspects of his experience.

Such absence of consistency on the part of the counselor sets limits to his empathic understanding since he is unable to be honest and direct in what he conveys. Lack of honesty and directness is also suggested for the inconsistent client who, like the counselor, feels threatened by the counseling situation and consequently wraps himself up in a tightly protective, uncommunicative package.

In order to determine the client's level of consistency or congruence the counselor listens to the client's words and observes his posture, gestures, eyes, and tone of voice. Then the counselor brings to the client's attention the contradictions in what the client says he feels and what the counselor notes the client's nonverbal language as "saying."

1

MS. B.: You say that you are happy about your new job but you clench your fists and your eyes seem to tell me you're hurting.

2

DR. T.: I wonder whether you know that whenever you talk about how wonderful your mother is, your lips begin to quiver and your brow wrinkles.

• • •

On some occasions the counselor may go even farther and express his own feelings about the client's behavior. He may say, "I know you just said you want to 'let yourself go' to enjoy life . . . but then you lean back in your chair and hold onto the arms.

I wonder? I feel uncomfortable because you are so sad. Yet, I get the feeling that you won't let yourself *feel* any other feelings."

When the counselor is forthright in his own feelings about the client's responses, he risks more than when he merely reflects the incongruence in what the client says and does. He not only risks being incorrect in his reflection but also risks exposure of his personal feelings. In other words, the counselor makes himself vulnerable to an attack on his feelings. Only the experienced counselor, well grounded in theory and nourished by supervised practice, who is congruent himself can successfully share experiences with the client.

Since reflection calls for highly developed skills, the beginning counselor is more apt to commit errors in reflecting. These errors relate to the more mechanical problem associated with timing as well as to the more complicated results of misunderstanding the client's meaning.

Errors in reflecting

Five forms of errors of reflecting are considered. Since they are interrelated, the beginning counselor frequently exhibits more than one.

Timing

The error of timing, which on the surface appears simple to avoid, is a constant difficulty. The greatest barrier to learning the "when" of reflecting stems from the lack of development of other skills such as attending to and understanding what the client says in addition to identifying feelings.

The essential generality about timing is that the counselor should not wait too long to reflect the client's feelings or the client may become confused by too many diverse feelings. At times, therefore, the counselor may need to interrupt the client's flow of words in order to focus on some important feelings.

Aside from this basic generality, the timing of reflection is also dependent on the counselor's empathic understanding of the client's life-style, of the meaning of the client's words, of the significance of the client's actions, and of the underlying feelings. The alert counselor makes certain that the client is ready for a comment. Readiness is not solely determined by when the client stops talking but also by when the client has con-

veyed a feeling that seems to need clarification.

The following two samples of client-counselor conversation contrast suitable and unsuitable timing.

1

ROGER: Wow, you're a great counselor! Man, no one has been able to get to me as well as you do. Really want to stop speeding up [use of methedrine, a form of amphetamine]. I feel great! Used to get all uptight and. . . .

MR. J.: Hold it, man, you're laying it on thick. Let's put it on the table. Feel good, want to stop. Swell, man, swell. Hearing something else though, coming from your drawn eyes, your pulled-in body, your tightly clenched arms. Level with me.

2

SEPH: You don't understand, how could you? Pig-lover, in with the cops, whaddya know about the street? That's where it's at—slaughter jungle. Nowhere, man, nowhere. No bread [money]—pieces of living. Gotta fox [woman]. Coke [cocaine] . . . sick with no dope. Whaddya know?

DR. K.: It seems to me that you are jumping to conclusions. I *do* understand. You believe all people are the same. You're angry, very angry. Isn't it time you looked at all this another way?

• • •

It should be noted that Dr. K. is missing the intense feeling of frustration and depression Seph is expressing. Seph's comments result in a standoff that permits him to avoid full responsibility for his actions. Since there are elements of truth in what Seph is saying, Dr. K. misses the meaning by calling attention only to the anger expressed. It would have been more productive timing if Dr. K. had stopped Seph's rambling at an earlier point when he was revealing the hopelessness of his relationship with Dr. K. An honest admission by Dr. K. of his feeling of discomfort and his lack of knowledge of the street would have placed both Seph and Dr. K. on a more even relationship. Another essential feeling that Dr. K. missed was Seph's remark that he is a street addict. This is a crucial factor since the street addict often relates more readily to his peers than to authority figures, as represented by Dr. K.

Worn-thin phrases

The frequency of timing errors is surpassed only by the rut some counselors dig themselves by means of monotonously using the same canned phrase to begin their remarks. "You feel" is the worst offender in this respect. The beginning of the counselor's remarks is worn thin by overuse. When the counselor regularly initiates his comments reflecting feelings with any one phrase, the result is either a client's deaf ear or an annoyed comment. These routine starters are used automatically. The very fact that they are automatic detracts from their impact and suggests that the counselor is doing too little "headwork."

Many other introductory phrases are available to provide some variation. Some of these are "You surely are . . . ," "It seems that you feel . . . ," "You believe . . . ," "Do I hear you say you feel . . . ?" "As I get it, you feel that . . . ," "It sounds like . . . ," "In other words, you feel . . . ," ". . . is that it?" and "Mmmhmmmm, I see; you're saying. . . ."

In addition to these openers, the counselor may come directly to the point of the feeling expressed by labeling the emotion.

DELORES: If I have to clean up after these kids one more time, I think I'll kill myself. Need some time alone. Someone older than 5 to talk to. Blood pressure is soaring . . . tired of it all.

MS. F.: Uptight, frustrated, depressed . . . that's what I hear you saying. Let's get into this. Tell me more about what you would rather be doing.

Another possible counselor response to Delores might have been, "I gather that you've had it. You sound frustrated, depressed, and lonely."

Too deep or too shallow responses

Improper timing and worn-thin phrases may occur more frequently but are less difficult errors to overcome than the counselor's lack of sensitivity to the depth of the client's feelings. When the counselor loads his response to the client's remarks too heavily (too deep) or tends not to go far enough in his response (too shallow), he reveals his lack

of understanding and/or inadequate expressive skills. One factor that may result in an inappropriate counselor response might be the counselor's level of regard for the client. The level of regard refers to the degree of respect and positive feelings (affective responses) that one individual has for another.

The counselor's reflection would most likely be inaccurate if his feelings toward the client were negative, too positive, or even neutral. Negative feelings arise when the counselor finds the client unappealing, distasteful, or objectionable. Neutral responses occur when the client just leaves the counselor cold and uninvolved. Too positive responses are just as serious since the counselor is apt to become too involved and overidentifies with the client. Either extreme, negative or positive regard, or neutral regard contributes to too deep or too shallow a reflection of feeling and/or experiences. Furthermore, the counselor whose regard varies with the client's moods is likely to communicate his changing feelings through too deep or too shallow reflections.

A consistent level of regard that leans toward positive feelings for the client encourages more accurate reflections. Adequate regard, empathic understanding, and satisfactory expressive skills turn out reflections that neither go beyond what the client intends (too deep) nor are less than the client implies (too shallow). The following examples show the contrast between too deep and too shallow reflection.

CHUCK: Where do I go from here? Wish I were different. Want to get into law yet don't feel in the bookgrind mood.

1

MS. O.: I hear you say you're feeling that you want to be a lawyer but you would rather not study. (TOO SHALLOW)

2

MS. O.: Chuck, I hear you say you're distressed because you are different, confused by your conflict. You want to study law but feel antagonistic toward the vigor of books. (TOO DEEP)

3

MS. O.: Chuck, I hear you say you feel uncertain about whether you really want to get into the book scene . . . yet you are interested in law. (PREFERABLE RESPONSE)

Selectivity

The error of improperly selecting the prominent feelings in the client's problem results primarily from at least three deficiencies of the counselor—a low level of observational and listening skills and, particularly, a lack of empathic understanding. Some psychologists explain selectivity in terms of an individual's basic attitudes toward himself and the people around him. An individual sees events and people not as they are but as he thinks they should be. This individual narrowing of the perceptual field is one of the factors in selectivity.

Every person is selective in his attention. No person sees, hears, or feels all of the sensory impressions around him. People are conditioned to sift out certain stimulations and attend to others. Otherwise the bombardment would be overwhelming. The effective counselor is aware of this selectivity and works harder at observing and listening.

When the counselor selectively attends to a client's feelings, he helps the client explore his emotional responses in order to increase the client's self-understanding. This procedure is less useful if the counselor is not tuned in to the client's concerns and instead directs the client to examine feelings that are too threatening for him at the present or, perhaps, not significant to the client's immediate problem. Ms. T. not only shows her bias but also her unsuitable and untimely reflection in her response to Gus in the following sample.

GUS: Ms. T., I've chosen to come to see you because you are a woman, and I thought you might be better able to clear up some of my confusions and conflicts. I'm married, have been for 20 years. For 18 of those 20 years my wife has taken care of the house, our two children, and has been active in community groups. Okay, along comes women's lib. Two years ago my wife joined an awareness group and since then talks all the time about how she has been locked or "thinged" as a household sex object. Ms. T., it's not that I want to keep her back, never did. She decided to stay home. But she's making me feel so defensive. So guilty that I can't relax. Even my

sexual responses are beginning to suffer; think she might consider me too aggressive. It bothers me so much; sometimes I'm even impotent. Whew! Said a lot.

1

MS. T.: Yes, indeed, I do understand you. You feel that the awareness group is competing with you for your wife's attention. You sound as if you're depressed. (UNSUITABLE REFLECTION)

2

MS. T.: Mmmmmm. I hear you telling me several things—pressures, feelings of guilt, beginning to restrain you from expressing yourself sexually. (MORE SUITABLE REFLECTION)

• • •

Close inspection of Ms. T.'s unsuitable response shows introductory remarks that seem worn thin ("you feel," "you sound"). Ms. T. also times her comment about the awareness group inappropriately. Gus's more significant concern appears to be his apparent loss of virility (impotence); he seems to feel pushed and is pushing himself into a passive role. Ms. T.'s shallow remark inaccurately selects certain parts of Gus's statement that are less essential at this counseling moment. Another error may be related to Ms. T.'s own feelings about women's liberation. Possible bias is revealed when she uses the word "competition" to explain Gus's reaction.

Another dimension of Gus's concerns is Ms. T.'s mention of Gus's depression. Gus may be depressed, although his comment does not state this. Depression does not seem to be the problem most important to him at the moment. Gus is asking Ms. T. for some reassurance in this crisis that is almost immobilizing him. Ms. T. must provide this reassurance by what she says and how she looks when she makes her remarks. Ms. T.'s unsuitable remark moves the discussion into another error of reflection.

Adding or subtracting

The counselor who tacks a thought or feeling onto the client's comments may be evidencing lack of skill, an attempt to exhibit his counseling power, or several other failings. This error in reflection is similar to the too deep reflective remark. Subtracting from the client's meaning or feelings is similar to the too shallow reflective expression. The counselor does not understand or does not listen carefully and so misses something important in what the client is saying.

The dissimilarity between adding/subtracting and too deep/too shallow remarks comes from the fact that in adding/subtracting the actual meaning of the client's statement is distorted. In too deep/too shallow remarks the meaning is not lost but the feelings are not responded to appropriately.

DONNA: I just can't see myself as a librarian all my life. It makes me feel hemmed in. I need a job where I move around and see people.

1

MR. I.: You don't think you could be tied down to the clerical work, the routines, the mechanical aspect of library work. What you need is a fast-moving job so you won't feel so restricted. Satisfaction of your need for people is vital to you. (ADDING)

2

MR. I.: You just don't like being a librarian. (SUBTRACTING)

3

MR. I.: I hear you saying you're feeling so uptight you don't know whether you can stand it much longer. You seem to want to break loose so you may meet more people. (TOO DEEP)

4

MR. I.: You feel you want a new job so you will be around people. (TOO SHALLOW)

5

MR. I.: Want to stop feeling hemmed in. You really are ready to start thinking about a new job, one that will have more to do with people. (MORE SUITABLE RESPONSE)

Exercise 7-2
ERRORS IN REFLECTING

Write counselor responses for the following client remarks. For client remark No. 1

show unsuitable selectivity, subtracting, and too shallow responses; for client remark No. 2, worn-thin phrases, adding, and too deep responses; and for client remark No. 3, inappropriate timing, unsuitable selectivity, and too deep responses. Also write a suitable response to each client. After you have completed writing your responses, discuss them in your larger group.

1

"I cannot help but pity my child. Twelve years old and he can hardly make himself understood; he has cerebral palsy [motor disturbance due to nonprogressive damage to the brain]. I'm sad about him, yet so ashamed—want to hide him."

2

"The nighttime is when it's worse. I have the same recurring nightmare—being choked. Wake up suddenly in a sweat, frightened, panic."

3

"Something I can't seem to help. Walk along the street and eye every man. Sex is my joy, the only act that turns me on. Think about it all of the time."

CONFRONTATION
Explanation of confrontation
Definition

The term "confrontation" has the tint of antagonism. Yet to confront suggests a much broader range of feelings. When the word "confront" is traced to its Latin roots, *con* is found to come from *com*, meaning "together" and *front* from *frons*, meaning "forehead" or "front." The original meaning of the word is to face, stand, or meet face to face. There are also three other meanings: to face or oppose boldly, defiantly, or antagonistically; to bring face to face; and to set side by side to compare. From these various definitions the concept of confrontation acquires negative hints of antagonism, positive glimmers of honest communication, and suggestions of objective gathering of facts for comparisons.

All of these definitions are used when one considers confrontation as an expressive skill. As an expressive skill, confrontation implies one additional thought; that is, the counselor confronts the client in order to close a gap that the counselor feels is separating the client from deeper self-exploration or from the counselor as a participant in the counseling session.

As an expressive skill, confrontation is a form of communication that may be defined as a process of calling attention to and/or challenging another person. This act brings the client face to face with the following:

- Discrepancies between the counselor's observations of the client's behavior and the client's statements about his experiences and his feelings
- Some facts the client does not know about himself or experiences only vaguely
- Something the client knows but thinks others don't know

The counselor who functions at higher levels often confines his comments (confrontations) to the client's adequacies and resources rather than to the client's shortcomings. Furthermore, confrontation is a reciprocal process insofar as the counselor makes himself vulnerable to the client's counterconfrontation. Thus both the client and the counselor may become unmasked by means of the tactic of "undenial"—confrontation that "tells it like it is."

No matter how constructive the confrontation, it still has a tendency to cause distress; the counselor must have well-developed empathic understanding so that he can assess what should be uncovered, when this recognition should be revealed, and how the screen of denial, aversion, and deception should be removed. Surprise, humor, or sometimes a forceful manner become channels by which the client's awareness is sharpened. Confrontation may be approached in a light vein with a simple question, "I wonder?" or may tend toward a direct frontal attack such as "Cut the bull!"

Forms

Confrontation pulls up short the confronted person. Some act of the confronter, of which he may or may not be aware, is a stimulus for the other person. This act of confrontation has many forms, which may either promote growth or be destructive.

Confrontation may vary from a light challenge to the client to examine and mobilize his resources for deeper self-recognition or constructive action in his own behalf to a thrust that may temporarily disturb the client's personal and social equilibrium. Daytop, Synanon, and some of the marathon groups are more likely to replace tact with brutal frankness. Although the "hot seat" is not comfortable, punishment is not the primary intention of the attack. Instead, in this electric group no holds are barred, and everyone undergoes the same penetrating and attacking scrutiny. Each individual is brought face to face with what others think and feel about him.

Communication may become more or less confronting, depending on the people involved; the words, gestures, and tone of voice used; and the time when the particular remarks are made. In addition, confrontation may be favorable and pleasant (positive) or unfavorable and repugnant (negative); it may respect the individual for his humanness (unconditional) or for what he does (conditional); and it is a symbol of one human touching another verbally or by means of nonverbal language (stroking).

General purposes of confronting

Long ago in Plato's "Apology," Socrates was quoted as follows: ". . . I say again that daily to discourse about virtue and of those other things about which you hear me examining myself and others, is the greatest good of man and that the unexamined life is not worth living." Synanon adds another thought to the purpose of confrontation. When what a person says he is doing does not show in his behavior, Synanon challenges this person with the criticism that he is "talking the talk but not walking the walk." Embedded in the two preceding quotes, which are centuries apart, are the purposes of confrontation.

Self-exploration

The essential purpose of confrontation is to produce suitable circumstances that encourage an individual to look at and hopefully to find ways to change his behavior. Confrontation is an invitation to someone to examine himself and to reflect on his behavior. The primary objective of confrontation is to free the confronted individual rather than to restrict, punish, or destroy him.

In order to avert increased self-depreciation, comments about weaknesses or limitations should be accompanied by comments about strengths or resources. Telling a person "You talk too much" can be devastating if the statement is not combined with "You have important ideas to contribute to the discussion. However, you talk so much that others don't have a chance for their contributions."

Self-exploration may arise from some feelings of discomfort by the individual or from the remarks of others. In either situation the value of confrontation comes from its immediacy and pertinence to the person's problems.

When someone else originates the confrontation he may say, "It seems to me that. . . . I wonder how it seems to you?" Whenever the counselor confronts his client, he controls the event by means of his selection. Because of his empathic understanding of the client, he is alert to what the client is ready to hear and what would be better omitted. The counselor confronts the client to focus his attention on a situation, not to mold him into a preconceived notion of what is right for him.

When the client attempts self-exploration, he turns to the counselor, with whom he feels he may safely explore himself and his difficulties. The client may need a relationship of trust and honesty that he feels he does not experience in his daily activities. In such an atmosphere the client discharges his accumulated frustrations and begins to confront himself with his role and responsibility in these frustrations. Whether these revelations become an active surge toward discovery of routes for behavior change or result in inhibiting the client depends on the counselor's skill in the give and take of confrontation.

Confrontation that leads to self-exploration is, in essence, a process of reality testing, with the ultimate goals of more effective intrapersonal (self-satisfying) and interpersonal (socially satisfying) living. In order to move toward these goals it is necessary to help the client remove the screen through which he sees himself and his experiences. Freud considered this as forcing the unconscious into consciousness.

In some instances this self-exploration may require the realization that the supposed freedom of choice an individual believes he has is actually compulsive behavior. Mary, for example, who says that she will date whomever she pleases, no matter what race or religion or "whatever my mother doesn't like," may be just as tied down by trying to do the opposite of what her mother demands as if she were adhering closely to her mother's requests. The question for Mary to examine would be, "How does she want to restrict her freedom, by self-imposed compulsions or mother-imposed demands?" Mary is not told what to do; she is just confronted with the reality of her choice.

Another example of confrontation is of particular importance for the individual who is physically handicapped or even temporarily ill. In either event the individual must be confronted with his existing limitations. Then the individual explores for himself the ways in which he must learn to live within the limits of his handicap or illness as well as the possibilities for him *in spite of* his limitations. At an even higher level of confrontation the handicapped individual may be helped to realize that he is able to express himself even better *because of* his handicap, for he has gained a broader and deeper understanding of living.

Behavioral change

Self-exploration initiated by confrontation is the foundation for behavioral change. In order for behavioral change to occur, the individual must go beyond self-examination to self-challenge and/or challenge by others. Before behavioral change may come about, the individual must develop an awareness of the behavior that should be changed (self-exploration) and also the alternatives for change from which he may select. Confrontation provides the necessary challenge to the client. The counselor underlines how his experiencing of the client and the client's expression of his experience are dissimilar. He says to the client, "You say you're perfectly satisfied with your sexual relationships with your present husband. How, then, do you explain what you said after your recent 'fling' with your former husband?"

Self-exploration and behavioral change are forerunners to becoming oneself more fully. The counselor confronts the client with ideas or feelings of which the client is oblivious or that he is avoiding. By awakening the client to these hidden areas the counselor strives to initiate or to improve the relationship between the client and himself or to open pathways toward solutions of the client's problems. This encouragement to action plus the challenge to the client to integrate, to become one with his experiences, offers the client the opportunity to become more fully sincere and in contact with himself, his strengths, and his resources as well as his self-destructive behavior.

Becoming oneself more fully

There has been a continuous theme throughout this discussion on confrontation that differentiates confrontation from other skills of communication. This difference stems from the fact that confrontation acts more as an initiator of new directions for the client (an act) than as a response to the client (a reaction). Confrontation does more leading than following; it steers the client away from the self-fulfilling prophecy behavior that makes him become a loser, a helpless and pathetic failure. The counselor charges the client to consider, "How much failing would you avoid if you honestly told your boss he gives you too much work? Look at what you do. You take on more work to outdo the others because you think you have to prove that you are not a loser. Then you find you can't finish. So you feel that you are a failure, and you do fail to finish. So you've proved to yourself and to others—you are a loser. You're throwing your own boomerang. Do you see that?"

The counselor confronts the client in order to spur him on to remove his "front." He helps the client either accept himself as he is or to find a new self-image, a new way of behaving, that would be more satisfying. In this way the counselor seeks the client's increased awareness and acceptance of his identity as an outcome of confrontation.

The falsity that "game behavior" produces in interpersonal transactions is another target for confrontation. Alcoholics Anonymous speaks of the "stinkin' thinkin'" of those members who play the rationalizing game. Berne (1964) writes of the numerous social games in which individuals indulge. Although certain games are expected because of the conventions of social living, ⸜blind adherence to these rituals or game behavior that becomes a way of life interferes with fulfilling living.

Stopping game behavior

"Calling the other's behavior" in order to stop game playing can be a troublesome confrontation. If the effort to stop game interaction is not handled competently, calling attention to it may reinforce it and result in branching out into further games.

Forms of games. The games people play are endless and the ways to cope with these games vary.

The sick game. Clients may use "symptoms" to cover up their real intentions. The client may be so tired, nervous, upset, and misunderstood that he is unable to work. The client actually wishes to stop working and to live a slow, leisurely pace but is unable to do so because of certain obligations or certain feelings about the work ethic.

The sick game has many variations. A young child may say that his throat feels "sore" and so he can't attend school that day and at frequent intervals thereafter. A wife who has nightly "headaches" may be unable to understand why her husband is annoyed with her lack of sexual response. The man who "just can't find the woman who could take care of him" may actually find most women annoying. Each of these individuals hides his game under certain conventions; the counselor must listen carefully to pick up the subtle messages that conceal the problems. Individuals play games in order to manipulate the people around them so that they may accomplish their goals.

The helpless game. Many individuals indulge in the pastime of helplessness for their own amusement or to score points with someone. Students ask questions for which they know the answers. They question because they have "psyched" the professor and know he enjoys the superior status of answerer. This kind of game is not serious unless it becomes the primary way in which an individual relates to other people. At times the helpless game becomes involved in such serious phobic behavior (excessive and unreasonably overwhelming fears) that the victim becomes unable to remain alone in a room or to walk out of doors alone. The client does indeed become a prisoner of his own symptoms.

The insight game. This game is much more subtle and possibly more manipulative than the sick or the helpless game. On the surface it appears that the individual is seeking better understanding of his problems. He keeps searching for greater insight. However, his search is endless, for he is actually defending his present behavior by means of his quest. He appears to be making progress as he grasps each new insight; yet this is just an illusion to keep him safe from change. The insight game is similar to intellectualization and often is associated with emotional insulation.

• • •

There are many more games, some of them everyday social events and others barriers to effective living. Some games should not be changed, for the individual has such a strong inclination for his game behavior that without his games he would function at a lower level. On the other hand, the three games just mentioned are often stumbling blocks within the counseling situation that detract from the client's ability to move toward self-fulfillment. Skillful, empathic confrontation at a suitable time may draw the game player toward more constructive behavior.

Game coping. There is a risk in confrontation, but without it there may be no further growth for the client. The risk is increased even more if an untrained, inexperienced, and therefore unskilled individual attempts confrontation.

Knowing the game one is playing does not in itself make it possible to change. After the counselor confronts the client with his game behavior, alternative behaviors must be offered for change. Withdrawal of reinforcement, calling the game, and the asocial response (Beier, 1966) are three interrelated methods of coping with game behavior.

The client whose behavior becomes bizarre may be playing a particular type of the sick game. Ignoring him when he puts his slacks on inside out or when he is unwashed and/or does not comb his hair is a form of withdrawal of reinforcement. He is also ignored when he jumbles words at high speed so that he cannot be understood. By ignoring the client's inappropriate behavior (negative reinforcement) and attending to the client's appropriate behavior (positive reinforcement) the counselor attempts to weaken and eventually to extinguish the socially destructive behavior (p. 68).

The counselor calls the client's game when, instead of ignoring, he brings certain inconsistencies to the client's attention.

1

The client implies that someone else must change before he can solve his problem.

TOM: If only my wife wouldn't make so many demands on me, I wouldn't drink.

MR. O.: Ten years you've been drinking, Tom. You've been married for 5. How come you drank before you were married?

2

The client suggests that past events force her to act as she does.

WILMA: My mother was just like that. I know I'm like her. She and my father would argue about such petty things as how high or how low the window shade should be. Then she would get more and more upset . . . throw up . . . my father would shut up. I'm just like her.

DR. N.: Are you saying that you can't be different just because of your mother?

• • •

In both instances the counselor strives to confront the client by interrupting the associations the client brings forth in support of his game behavior. This is not done to condemn the client but rather to confront him with new ways of looking at his behavior.

Beier (1966) writes of the "beneficial uncertainty" he establishes when he removes himself from the client's game with an asocial statement. This asocial statement is a shocker that brings the client up short. The client is uncertain about what the counselor is trying to accomplish. However, in the beneficial atmosphere of the acceptant, nonthreatening counseling session, the client is more likely to attain greater freedom of response. The asocial response does not fit into the usual rituals required by conventional conversation and the client must do a fast turnabout to a new trend of discussion.

1

TED: You're a jerk. You don't understand anything.

MR. R.: Aren't you foolish talking to a jerk who doesn't understand?

2

MARTHA: Heh! Wait a minute. I think you're manipulating me.

MR. T.: So?

• • •

Underlying the counselor's remarks is a challenge to encourage the client to spread the nature of his interactions. In the supportive counselor-client relationship the counselor confronts the client with the offbeat remark so that the client must rise to new ways of relating. This asocial response also has the effect of inducing cognitive dissonance.

Reducing cognitive dissonance

The word "cognitive" comes from the Latin *cog-nascere*, meaning "to get to know"; "dissonance" stems from the Latin *dissonant*, meaning "disagrees in sound" and therefore out of harmony. Hence cognitive dissonance refers to the process of receiving information (getting to know) that is not harmonious (dissonant) with one's already existing beliefs and/or knowledge. Confrontation induces this state of conflict when beliefs or assumptions are challenged by the contradictory information the counselor presents to the client.

When this dissonance occurs, the client feels uncomfortable and seeks to correct this difference between what he believes and the new information presented to him by convincing himself that the differences do not exist (denial or intellectualization), by adopting some other kind of defensive coping strategies (rationalizing), or by controlling the flow of information (withdrawal or fantasizing). It is the counselor's responsibility to help the client over this distress of cognitive dissonance so that the experience is helpful rather than destructive. One way to do this is to suggest new actions that contradict the negative attitudes.

The assumption underlying the promotion of new actions is that changing behavior induces a change in attitudes. The individual who complains he is unable to speak on the telephone because he stutters is urged to make phone calls. Although the act of phoning does not cure his stuttering, the successful act, if repeated, gradually decreases the withdrawal attitude.

It is up to the counselor to confront the client with the facts that he distorts or screens out so that he cannot continue to maintain and protect his misbeliefs. After this awareness the client is assisted to engage in the behavior he has avoided because it has created negative attitudes. When the individual commits himself to certain behavior, he also makes an effort to reduce the dissonance. Gradually the negative attitudes change and the individual decides that the behavior is not so bad after all.

Example 7-1 describes Bill, who committed himself to a change of behavior and in turn changed his attitudes toward himself.

Example 7-1
BILL MAKES THE CHANGE

A brief examination of Bill exemplifies how commitment altered both attitudes and certain self-destructive behavior.

Bill grew up in an affluent middle-class suburb. Through high school he remained a conformist to the customs of the establishment. He wore his hair clipped short and dressed in the prescribed slacks with a belt, shoes below the ankle bone, and socks to match. He was, as he later stated, "a real square." He graduated from high school in 1966, wearing the traditional gown with the mortar board correctly placed on his still closely cut hair. He also bit his nails.

He went off to the conservative college he had selected. During his first year at college, away from the close scrutiny of his liberal, yet "square," parents, he encountered some new ideas. He discovered that smoking pot was not so bad. He found that the dudes with long hair whom he had avoided during high school were interesting and pleasant friends. By the end of his first year at college his hair was longer and he wore faded jeans and a tie-dyed tee shirt. Gradually his nonpolitical approach to life was disturbed by some of the injustices he heard about. The Vietnam War rattled his calm, yet his negative attitude toward participating in mass sit-downs continued. He also continued to bite his nails.

During his second year at college he joined a sensitivity group and met Raina. In the sensitivity group he voiced his anxieties and was confronted with disagreements, with a radical political philosophy as well as an acceptant, supportive atmosphere for change. He became aware of different viewpoints and very much aware of Raina.

Raina's life-style was a free one. She reveled in experiences. She liked to travel. She enjoyed and was comfortable with people. She was distrustful of the way in which the government functioned. Bill and Raina had spritely conversations about their differing viewpoints. Bill began to wonder and still bit his nails.

The third year of college was a year of decision. Bill went on a trip and left his razor at college. After the itchy period of the newly sprouting face hair, a shaggy beard and moustache began to grow. The hair spread and by the time he went home his parents were confronted by a shaggy-haired, bearded, tall young man whose ideas were more radical than theirs. He spoke of the blacks and the rip-off they experienced. He condemned the administration for the war. He joined demonstrations against the war. Bill had advanced from cognitive dissonance to a commitment, to behavior change. He was trying to let his nails grow.

College graduation was a hassle. Bill was pleased that he was graduating but resented the requirement that he wear a cap and gown. He decided to join several others who were protesting by sitting in the audience rather than participating in the ceremonies. He asked his parents to join him in the audience, and they complied with his wishes. He had stopped biting his nails for a few months but recently had begun to bite them again.

After graduation, Bill drove to the West Coast with some of his friends. They stopped at the homes of various friends, crash-padding along the way. He enjoyed the freedom of the trip and the company of his friends as they camped under the vast expanses of the open spaces. The stares of people at the shops where they bought their supplies did not disturb him as much as the reaction of the cousin of one of his friends at whose home they were supposed to stay overnight. The cousin asked them to leave because her husband hated long-haired hippie types. However, the support of his friends and their mutual annoyance made the twinges of distress less painful. He knew he would be likely to bite his nails.

Bill decided to move to the "haven of culture" and to his "kind of people" in Boston. He wanted a year of work before he decided what he would do next. Raina was still in his thoughts, but by this time they both had found other friends. Bill felt so different about everything. He moved into an apartment with some friends. Together they collected furnishings from Goodwill, The Salvation Army, parents, and other friends. He decided that he liked where he was at—that he liked what he was doing. He thought he would like to become a poverty lawyer. He decided to stop biting his nails and did just that.

INTERPRETATION

Interpretation is another form of an expressive skill. The skill involved is more difficult to attain since interpretation requires the counselor to color his remarks with his own way of thinking. Thus the counselor introduces his own frame of reference into the counseling session.

The challenge to the counselor is immense. The counselor must state his view of the client's beliefs in language that the client can understand and hopefully accept. Therefore the counselor must talk the client's talk even though he adds a thought or views an idea somewhat differently.

Some counselors and psychotherapists cloak their interpretations in complex explanations about how the past explains the present. This grows out of the notion that the client's history accounts for his present story. These interpretations require extensive knowledge of symbolic formulas that link the past with the present.

Other counselors and psychotherapists seek interpretations based on the client's present behavior and the way in which he sees the past in the here and now. The characteristics that tie all interpretations together are empathy, creativity, and well-developed listening skills.

Explanation of interpretation

Interpretation brings new facts or rearranges existing facts so that the client may see his behavior in a different light.

Definition

Interpretation is a suggestion that goes a little farther than the client's suggestions about his ideas and feelings. Interpretation is a contrast in which the counselor mixes together two or more of the client's dissimilar ideas.

A broad, general definition of interpretation is the slanting of the client's comments from another standpoint. This slant provides the basis for the client to view his concerns with a fresh start. The counselor binds together the client's words and feelings with his altered choice of words, uniting these words with additional relevant information. Thus interpretation differs from:

- Paraphrasing, which focuses on repeating, highlighting, or coordinating of the content of the client's immediate conversation
- Summarization, which repeats, highlights, and coordinates the content and feelings of the client's conversation over an entire session or a series of sessions
- Reflection, which is the act of bringing to the client's attention the feelings he is expressing
- Confrontation, which brings the client face to face with some gaps in what he is saying and/or some facts that may be vague or unknown to him

Interpretation contains all of the characteristics of the previously described expressive skills as well as something else. This something else is the particular meaning or significance added to the client's discussions. For instance, the following samples of behavior and counselor's different responses compare these expressive skills.

STUDENT: I feel terrible about being absent from class so much.

1

INSTRUCTOR: You've missed several classes. (PARAPHRASE)

2

INSTRUCTOR: You sound upset and worried. (REFLECTION OF FEELING)

3

INSTRUCTOR: Yes. You have had so many absences that your grade is very likely to be affected. What do you plan to do about it? (CONFRONTATION)

4

INSTRUCTOR: You've missed so many classes . . . now you're worried about your grades. Have you thought about the possibility

that you may actually be setting yourself to fail? (INTERPRETATION)

• • •

In each response there are elements of observation, description of behavior, and explanation. In general the primary purpose of interpretation would be to provide the client with a new look at himself and his problems.

Aims

The working hypotheses present some guesses about the significance of the client's behavior so that the client may coordinate his thoughts and feelings. In the process of coordinating his thoughts and feelings the client attains a new outlook (reconstruction) of the meaning of his behavior. Finally, out of the original hypothesis, the coordination, and the reconstruction comes about an expansion of awareness and self-realization.

Hypothesis. The counselor explains his guesswork, or hypothesis, about the client's problems according to the counselor's orientation. If insight is the counselor's eventual goal for the client, then the counselor seeks to make conscious the hidden meanings of the client's behavior patterns. The counselor would guess about (hypothesize) the underlying conflicts and instinctual needs driving the client toward certain forms of behavior.

If reduction of symptoms and changed behavior are the primary goals, then the counselor sets forth his hypothesis about the inappropriateness or appropriateness of the client's behavior. The client follows through by testing the counselor's hypothesis about his behavior by trying out new ways of behaving both in and out of the counseling session.

The counselor in either an insight-oriented or behavior-oriented approach states his hypothesis in words familiar to the client. Thereafter the hypothesis becomes the signal that prods the client to work through his problems. Putting together the hypothesis and the changes in behavior is, of course, dependent on the client accepting the counselor's guesswork. The accepted hypothesis, in this way, becomes the building block for coordination.

Coordination. The counselor assembles his observations of the client, the client's comments about his feelings and about other people, and his own (the counselor's) frame of reference. This coordination of observation, feelings, and frame of reference is submitted to the client for him to decide if it *fits* him.

By means of this coordination the counselor accomplishes at least two purposes. First, he emphasizes the similarities and differences of the client's and the counselor's views of the client's behavior. Second, he compares the client's past and his present behavior. Coordination of these various aspects hopefully clarifies the client's thoughts and feelings and help focus his attention on some important issues.

In one sense the counselor's coordination interferes with the client's ongoing behavior and leads the client into constructing new or reconstructing old ways of behavior. In the fourth response on p. 162, when the instructor coordinated the student's absences and his worry with the interpretation of the student setting himself to fail, the instructor was stating an hypothesis and was coordinating the behavior and the feelings into meaning. The student might respond in many ways; he might resist the interpretation (deny it) or begin to wonder about himself (reconstruction).

Reconstruction. As the client becomes able to talk more freely about his present as well as his earlier conditions of anxiety arousal, he also begins to display new responses to his present situation. Apparently the client is now able to handle daily affairs with a lower anxiety level. His reconstructed behavior (new responses) begins to appear outside the counseling office. As this occurs, the client moves toward termination of the counseling sessions.

Sometimes the client switches to resistance rather than to reconstruction. Sufficient self-reinforcement or reinforcement by others may act as a counterforce to the reconstructed behavior. For this reason the drug addict continues shooting up even though it is becoming more difficult to steal enough money to buy the increasing number of bags of dope. The sudden pleasant flush from the injection maintains the habit in spite of the counselor's hypothesis: "Has it ever occurred to you that this self-destruction has something to do with your guilt feelings?"

The resistant client finds it extremely difficult to give up his behavior even though it is inadequate and not completely satisfying. The client is reluctant to reconstruct his behavior because the promise of other reinforcement is insufficient for him to take the risk. The counselor's hypothesis and subsequent coordination do not fit the client. Example 7-2 describes the difficulties an individual has in restructuring his behavior even though he accepts the interviewer's interpretation.

Example 7-2
INTERPRETATION: BERNIE ACCEPTS, MARTHA REJECTS

Bernie returns to his home after a session with Dr. I. and says to his wife:

BERNIE: Martha, this is it. I *am* going to stop drinking. Doc has been saying it, now I know it. Alcoholic, I'm an alcoholic.

MARTHA: Yes, so what's new?

BERNIE: Martha, I'm trying to tell you something. Something it's taken almost 20 years to accept. I'm a souse, a lush.

MARTHA: Okay. So you drink too much. I've known that. What is that guy, Dr. L., trying to get you to do? Pulling you down this way. You oughtta stop seeing him. He's a quack!

BERNIE: But, Martha, that's not it—not it at all. In the group, and alone with Doc, I've been putting things together. How I've done things, not able to work, a weekend drunk. Many Mondays, no go, and you. . . .

MARTHA: Yeah, now you're going to blame me. I make you drink, don't I?

BERNIE: No, wasn't going to say that. But, Martha, Martha, listen. Remember when I stopped for a month? No piling it on, on weekends, for 1 month. Remember how mad you were when I wanted you to just sit and talk to me? To go out into the woods? Let loose? Once you even said I should get something to wet me down. You wanted to be left alone.

MARTHA: See. I thought so. It's that Dr. L.; he's trying to get you to hate me, to blame me, for your drinking.

BERNIE: No, Martha, that's not it. Oh, what the hell! *(Walks to the door swiftly, swings it open, slams it behind him, and leaves the house.)*

One might question whether Dr. L. has sufficiently helped Bernie to prepare for this attempted reconstruction of his behavior. It seems that Bernie accepts the hypothesis that he is an alcoholic. There are numerous facts to support this guess. Also, he is ready to coordinate his feelings about himself, about his wife, and about his relationship with his wife. He is even ready to do something about changing his behavior. Every item fits except for the way he must learn to deal with his wife. The interpretation was inadequate since it did not help Bernie determine how he might handle himself outside of the counseling situation.

Martha interpreted also. She interpreted Dr. L.'s speculations as a threat to her. She may also have been reluctant to release Bernie from his dependent role. She preferred reinforcing the dependence that alcohol encouraged in Bernie. This newfound strength of Bernie's to change was not to her liking. Bernie needs to know more about her interpretation.

Bernie is ready to change. He is prepared to become more self-sufficient and less dependent on Martha. The counselor has brought Bernie to this stage by means of the hypotheses, which pointed out the reinforcements for Bernie's continuing alcoholism. However, Martha places hurdles before Bernie. If Bernie is sober, he wants Martha's companionship. If Bernie doesn't drink, he is more capable of handling his own affairs. Martha interprets this as a form of rejection. She is reinforced by Bernie's weakness.

Insofar as Bernie is concerned, the goal of independence looks farther away. An additional negative outcome might be Martha's rejection of him if he stops drinking. Martha reinforces the goal of "sickness," rather than "wellness."

The client may have to be exposed to numerous interpretations of his behavior and the behavior of those around him. He must be prepared for the outcomes of his behavior change (reconstruction) on the relationships involved as well as for the expanding realization of his personal growth.

Expanding boundaries. Broadening self-knowledge makes it easier to deal with the "shoulds" and the "should nots" that are often hurdles in the way to fulfillment. The person who does mostly what he should do when he should do it is not free enough and usually not aware enough to change his behavior. Interpretation may open new doors to experience and change for the client.

The client who can look at himself through the eyes of the counselor and then become his own observer is on the way to experimenting with new behaviors. At first the client must be amply reinforced, with sufficient time to try out alternative roles. Gradually he finds his own more effective role, his own pattern of behavior, his reconstructed life-style.

This broadened base of functioning may be compared with the completion of a puzzle in which one of the pieces is turned the wrong way or is missing. Someone, in this instance the counselor, turns the piece around or provides the missing piece. After this assistance, this interpretation, the client is able to put the puzzle together. The next time, or it may take one or more similar "pieces" of interpretation, the, client expands his awareness of the possible difficulties with his situational puzzle, and he takes over.

Another way of looking at this expanding awareness is to think of the client as someone who has developed an inadequate script for his behavior. This script contains conversations and hypotheses that the client has acquired from early infancy. The counselor points out the hypotheses of the client's script, shows him the way to compare, coordinate, and change, and reinforces him to reconstruct his behavior. The client becomes more informed about how to solve his problems and tries to get along on his own.

Interpretation may also be considered as a form of learning to discriminate. The client builds up freer, more effective functioning because he learns the difference between his own view of his behavior and the counselor's hypothesis about the client's behavior. As a result of these interpretations and differentiating reinforcements, the client breaks up his rigidity and establishes more positively active and flexible behavior.

TRUAX TRIAD:
empathy, genuineness, and nonpossessive warmth

Empathy, genuineness, and warmth. (Courtesy H. Armstrong Roberts, Inc.; from Poland, R. G.: Human experience: a psychology of growth, St. Louis, 1974, The C. V. Mosby Co.)

Alice (In Wonderland) asks:

"Would you tell me, please, which way I ought to go from here?"

"That depends a good deal on where you want to get to," said the Cat.

"I don't much care—" said Alice.

"Then it doesn't matter which way you go," said the Cat.

"—so long as I get *somewhere*," Alice added as an explanation.

"Oh, you're sure to do that," said the Cat, "if you only walk long enough."

• • •

Doubts and concerns about "Which way I ought to go" are often signals of becoming and of growth. The wise Cat knows that only the individual can set his own direction of where he wants to go or what he will become.

There are some signs to follow, certain attitudes, skills, and knowledge that can smooth the journey. Three necessary items for the journey are empathy, genuineness, and warmth. These three characteristics help the human service worker experience understanding, openness, honesty, and intimacy if he "only walks long enough."

EXPLANATION OF TERMS

Empathy, genuineness, and warmth are basic to the effective interview, to the counseling session, and to psychotherapy. The importance of these characteristics is supported by the findings from many years of research. These characteristics are referred to as the Truax triad, from Truax and Carkhuff who first pulled together these three aspects in their book *Toward Effective Counseling and Psychotherapy* (1967).

An explanation of these terms starts with an examination of the interrelationships among these three aspects and a more specific probing of the definition, general characteristics, and scales for measuring each of these concepts.

Some research studies have indicated that empathy and warmth must go together for a successful helping transaction. Other studies support genuineness or warmth as the primary prerequisite to the interview. Which comes first is probably not as significant as the fact that either warmth or genuineness must be at a high level with empathy for an effective relationship to exist and persist.

The "halo" effect may also enter into the relationship of the three attributes. This effect occurs when an individual uses his first or initial impressions as the standard by which he judges whatever happens afterwards. For example, Melinda does it when she says, "I know it was going to be good. He was just out a sight when I first met him." It is what the voter is doing when he says, "I don't like his looks. He can't be trusted. I'm voting for . . . whose tax policy is good." In each instance the reacting person picks out some characteristic or impression and uses it to explain all other characteristics and impressions, as if the individual were enveloped in a "halo" that describes everything.

Other items also contribute to the successful outcome of the counseling transaction. These may be the depth of knowledge of human behavior, skill in applying behavior modification techniques, and additional interpersonal skills such as an internal rather than an external frame of reference. Thus the counselor views a problem from the uniqueness of the client's position (internal) rather than from the position of an outsider (external).

It may be said that although other conditions are very likely to be necessary to conduct an effective helping transaction, empathy, genuineness, and warmth are the basic attributes. The order of importance of these attributes is difficult to establish, but there are indications that the sequence may be genuineness, warmth, and empathy—the counselor's work, which is the outcome of genuineness and warmth.

Empathy

Empathy is a special kind of observation, experiencing, and communication. The empathic counselor remains himself yet looks at the world through the client's eyes and listens with the client's ears. In other words, the counselor walks in the client's shoes, tries on the client's point of view, so that he may experience the client's feelings. After observing *as if* he were the client and experiencing *as if* he were the client, the counselor feeds back to the client what he, the counselor, has seen, heard, and felt. Empathy requires the counselor to personalize his communication in the client's language.

The counselor temporarily identifies with the client but must constantly be aware that he *is not* the client, but rather a participant-observer. It is important for the counselor to have both insight and outsight. Insight helps the counselor understand himself and outsight helps him to intellectually understand the client. Empathy goes one step further. Empathy starts with insight and adds outsight plus a feeling for the client's meanings and emotions. The empathic counselor goes beyond the client's public showing of his behavior to the internal world of the client.

A German psychologist invented the word *einfuhling* (*ein*, meaning "in"; *fuhling*, meaning "feeling") to describe the process of one person feeling himself into another person's private world. Feeling another's world is easier if one has had similar experiences. Perhaps this explains the deeper understanding of former alcoholics for other alcoholics, of former junkies for continuing drug addicts, and of people who live in the inner city for other inner-city dwellers. For example:

MARIO: How can you know what it's like to split from the scene? You're the cat who's made it.

DR. P.: Yep. You're right. I'm sitting in this foam-soft chair now, but I used to be on the street, the highly charged streets of Harlem, where verbal shoot-outs were just the beginning.

Definition

Empathy is the active, persistent, and respectful tuning in of one person to another's wavelength in order to receive and share the other person's meanings and feelings about his experiences. In this intense and intimate interpersonal process the counselor reaches out and walks hand in hand with the client. In one sense, empathy signifies that the counselor sensitively plays the client's role. This role-playing permits the counselor to experientially recognize and understand the client who feels rejected, persecuted, or mistreated or who acts irrationally.

General characteristics

Unconditional acceptance. Unconditional acceptance is built on the groundwork of genuineness and warmth. Sincerity and openness (genuineness) as well as comforting and tender feelings (warmth) are the essential characteristics of acceptance. Empathy

is both part of and the result of acceptance. One person cannot accept another unless he understands the values of the other person as he is, not as he ought to be.

Another way of looking at acceptance is to start at what must be the beginning—self-acceptance. Self-acceptance encourages genuineness and warmth, which lead to empathy. Empathy is actually acceptance of another individual Acceptance of (and by) others increases self-acceptance. Additional increases in any of the elements of the acceptance-empathy process will have a positive effect on the interpersonal relationship.

There are two additional characteristics of unconditional acceptance that are important to consider—brotherly love and altruism. Brotherly love is similar to the Greek word *agape*, which refers to nonsexual, unselfish caring for another human being. Brotherly love "means the sense of responsibility, care, respect, knowledge of any other human being, the wish to further his life. . . . In brotherly love there is the experience of union with all men, of human solidarity, of human at-onement" (Fromm, 1956).

Agape is intertwined with altruism. The altruistic person considers the welfare and happiness of another above his own welfare and happiness. This altruism encourages an individual to make possible another person's satisfactions and self-fulfillment. Unconditional acceptance is all of these aspects.

Openness in receiving and recognizing. Acceptance moves a person into an open relationship of receiving and recognizing. Nonacceptance of oneself and/or others tends to set up nonrelationships in which there is less self-disclosure.

Openness to different value systems, to the opposite sex, and to bizarre behavior and speech is an essential characteristic of the acceptant counselor. Such openness assists the counselor to develop a fuller contact with the client. The counselor then encourages and receives information more freely and recognizes the gaps in the information presented.

Openness helps the counselor listen more attentively, sensitively, and calmly. It keeps the counselor in touch with the ways in which the client is like him and the ways in which they differ. Finally, openness tells the client, "I am here with you." All these aspects are part of acceptance and empathy.

Accurate processing of information.
The sorting or processing of information is
one of the counselor's goal. He sets in order
his experiences of the client. The counselor
systematically and constantly sifts through,
groups together, and screens the feelings,
moods, and inclinations of the client. Out of
the assorted jumble the counselor files some
ideas about the client (see discussion of multiple hypotheses on pp. 36-39) and some approaches for alternative solutions to the
client's problems. At the appropriate moment the counselor shares these findings
with the client so that he, too, may sift,
group, and screen. Then the client shapes
his own words and images. Part of the client's
work is to use this new awareness to discover
what his purposes are, how he is fulfilling his
purposes, and how he may better fulfill his
purposes.

Concrete feedback. Concrete, moment-
to-moment, appropriate counselor feedback
is basic for the client to go forward in his
searching and becoming. This feedback is
directly related to the counselor's degree of
accuracy in processing the interpersonal relations between the client and himself. The
counselor "cools it" in his return message
(feedback) to the client. Although he is sensitive to the client's feelings and experiencing, he does not lose the balance between
being inside the client's world (empathic)
and yet sufficiently outside (objective) so
that he can be exact and compact in reporting to the client.

Another requirement for concrete feedback is the counselor's moment-to-moment
tryout of his impressions to make certain
that he thoroughly grasps the client's meaning. For example:

MS. R.: You seem to feel that it's the scar on
your face that keeps people away from
you. Is that what you're saying?

HENRIETTA: That's it, almost, but—well.
Oh, hell, if only people were color-blind.
White on the outside, black on the inside—
that's me.

MS. R.: I'm hearing something else now. Is it
that . . . are you really more upset because
of your parents? You said earlier you
wished you had "come out black"—would
have been easier.

Ms. R. has caught the inadequacy of her
interpretation of Henrietta's problem. Hen-

rietta is using her facial scar to conceal her
real distress, which appears to be her discomfort with looking and passing as a white
woman when her mother is black. Henrietta
seems to be telling the counselor that her
life would be simpler and less frustrating if
both her parents were black.

Moment-to-moment tryouts check the
counselor's errors before they get too far
off the beam. This immediacy of experiencing, reporting, and correcting allows for a
more direct, complete, and positive relationship. Immediacy also firms up the client-
counselor partnership and exploration of the
transactions that come about from their
relationship. The more empathic counselor
responses are the following:

- Those that fall into place with what's
happening in the client-counselor relationship
- Those that are responsive to the "good,"
"bad," and/or mixed-up feelings implied
by the client's discussion of an event
- Those that are more likely to concentrate
on deeper exploration of fewer topics

In essence the counselor unconditionally
accepts the client, is open to clearly receive
and recognize the client's feelings, and puts
together all that he understands about the
client. The counselor who is functioning at
a higher level also continuously checks the
accuracy of his findings by concretely and
specifically reporting them to the client. In
this way the counselor prepares and encourages the client to explore and disclose his
deeper feelings and experiences.

Empathy scale (Table 5)

The empathy scale serves as a means to
rate the counselor's functioning level of
empathy. This scale is adapted from several
sources (Barrett-Lennard, 1962; Carkhuff,
1969; Truax and Carkhuff, 1967). The characteristics (dimensions) included in the
empathy scale include the degree of the
counselor's awareness of feelings expressed,
the concreteness of the counselor's responses,
the immediacy of the relationship, and
the depth of self-exploration. The five
levels of the Truax-Carkhuff scale are re-
tained as well as the idea that level 3 is
the minimum facilitative level. A score
sheet with a method for determining the
counselor's functioning level follows the
rating scale.

Table 5. Scale for rating functioning level of empathy*

AWARENESS OF FEELING	CONCRETENESS	IMMEDIACY	SELF-EXPLORATION
LEVEL 1			
C. responds in a distant and unrelated manner. C. shows little awareness of even Cl.'s obvious feelings. C. may try to understand but from his own viewpoint.	C. leads discussion or responds to client in an unclear, non-specific, overintellectualized manner, avoiding personally significant situations or feelings.	C. deals with many items but does not attend to or speak of them in relation to the words and feelings of the client-counselor relationship even though Cl. refers to C. in his remarks.	Cl. mechanically talks about himself, his problems, or his feelings. C. fails to encourage Cl. to produce personal and/or emotional material. Cl., in effect, does not reveal himself, either because of lack of encouragement from C. or because he actively avoids discussing more personal concerns.
LEVEL 2			
C. shows poor understanding of the meaning of Cl.'s expressions. C. often responds accurately to Cl.'s obvious (surface) feelings but overlooks the depth of Cl.'s feelings. Sometimes C. jumps to the conclusion that Cl. feels more strongly or more concerned than he actually is.	C. may talk about Cl.'s feelings and experiences but the discussion is unclear, intellectualized, and not sufficiently specific.	C. seems to disregard most of Cl.'s verbal and nonverbal expressions that may refer to C.	Cl. often responds mechanically without exploring the meaning of experiences and does not attempt to unveil or understand feelings. When C. tries to encourage Cl. to discuss personally relevant materials, Cl. may agree or disagree, change the subject, or refuse to respond. Cl. does not produce new information related to his problems.
LEVEL 3 (minimum facilitative level)			
C. accurately responds with understanding of Cl.'s obvious feelings but does not realize how intensely Cl. feels about some of the material discussed. C. may misinterpret deeper feelings.	At times C. enables Cl. to directly discuss more personally significant material clearly and concisely. In some areas of discussion C. does not assist Cl. to be sufficiently specific.	C.'s verbal and nonverbal behavior is appropriate for Cl.'s comments about others and to some extent about himself (C.) in the counselor-client relationship but is not sufficiently clear in relation to the client-counselor relationship.	Some personally relevant and new material is willingly produced but discussed as if it has been rehearsed by Cl. Cl. shows some degree of either feeling or spontaneity but often not both responses.
LEVEL 4			
C. is sensitive to Cl.'s obvious and deeper feelings and accurately communicates his understanding to Cl., thus enabling Cl. to express feelings he was unable to talk about previously. C. tries to see things through Cl.'s eyes.	C. often helps Cl. discuss personally significant feelings and experiences in specific and concise terms.	C. openly, yet in someways cautiously, pulls in Cl.'s comments to refer to the client-counselor relationship.	Some personally relevant and new material is willingly introduced and openly discussed with emotional expressiveness. Cl.'s verbal and nonverbal behavior fits the feelings and the information discussed. C. begins to help

Continued.

Table 5. Scale for rating functioning level of empathy—cont'd

AWARENESS OF FEELING	CONCRETENESS	IMMEDIACY	SELF-EXPLORATION
LEVEL 4—cont'd			
			Cl. get deeper into his relationships with others (interpersonal), yet Cl. is not fully enabled to discuss these relationships.
LEVEL 5			
C. accurately responds to Cl.'s obvious and even his most painful feelings. C. is tuned in to Cl.'s deeper feelings but is not burdened or distressed by them. Together C. and Cl. explore feelings deeper than what Cl. was able to express previously. C. appreciates the meaning and importance of Cl.'s experiences.	C. enables Cl. to freely and fully discuss personally significant feelings and experiences in specific and concise terms regardless of the emotions expressed.	C. openly responds and directly pulls in and interprets Cl.'s comments to refer to the client-counselor relationship.	Cl. is able to be himself as well as to explore himself. Cl. actively and willingly engages in careful inward searching (intrapersonal) and discovers new views about himself, his feelings, and his experiences. Cl. arrives at some choices for behavior change based on his new view of others as well as of himself.

*C. = counselor; Cl. = client.

SCORE SHEET FOR EMPATHY SCALE

Place check under *one* of the levels for *each* characteristic. The empathy level equals the total number of checks under each category divided by 4, the total number of categories. This is the average level of all the characteristics.

EXAMPLE:

Feeling	Level 2
Concreteness	Level 3
Immediacy	Level 2
Self-exploration	Level 1
	Total = 8
	Divided by 4 = 2 (level 2)

Characteristics	Levels				
	1	2	3*	4	5
Awareness of feeling					
Concreteness					
Immediacy					
Self-exploration					
Total No. of checks under each category					

*Minimum facilitative level.

In the following exercises, two different approaches to raising the level of empathy are described. In Exercise 8-1 the trainee discovers his general approach to understanding the meaning of the client's experiences and to recognizing the client's feelings. The exercises that follow this global exercise focus on more specific aspects of empathic ability.

Exercise 8-1
TRYOUTS FOR EMPATHY

One way to improve the level of empathy is to role-play a variety of situations. The following procedure offers a global approach to discovering the functioning level of empathy and then assists in improving the level by means of feedback and repetition.

Get into groups of five and select two role-players and three observers. Each observer should have one copy of the scale and two score sheets.

The role-players select a situation from the three following situations and seat themselves so that the three observers will have a good view of them.

1

GERT: *(Sits with her legs tightly crossed and her hands clasped in her lap. A thin smile spreads across her face as she looks at Mr. L. and says in a high, shrill voice—)* But, I don't really care about the scar on my face—doesn't bother me at all. My concern is with college, 4-year college to which I should transfer after this semester.

2

PETER: I do okay on the job until the voices come. Then I stop, listen. They talk to me so I can't work. Once I caught my hand in the machine because the voices were talking.

3

MS. L.: Can't stand it any more, married to that, that. . . . Now he's joined the gay liberation. Can't go on—pretense, phony. Can't, can't! Hide it from my friends, hide it from my parents, hide—hide. At first, didn't matter—gentle, so gentle. We lived our own sex hangups, but now he flaunts his gay. . . . Oh, no, what should I do?

Exercise 8-2
SPEAKING TO THE POINT

Rate the following counselor responses according to the degree of concreteness on a scale from 1 to 5. Refer to the second column (concreteness) in Table 5 for characteristics at each level. After completing your ratings, compare your scoring with others in your group.

MAC: I find your explanation of my father's insistence on my going to medical school unacceptable. How can it be either/or?

Possible replies by Mr. J., the counselor:
- "I see. So you don't believe it can be either/or?"
- "You know you don't have to agree with me."
- "I see what you mean. Maybe you can tell me more about how you see it so I might better understand how you feel."
- "Well, perhaps it's because you really don't want to understand."
- "You mean that you can't see your father's viewpoint?"
- "Hmmmm. Lots of sons find it difficult to understand how their fathers feel."
- "You know, you've got something there. Let's look at this some more."
- "Yes, I did make a broad statement. Just because three generations have been doctors doesn't mean that either you have to be one or leave home. Hmmm, let's see."
- "I understand your feeling. Feel hemmed in—is that it? Maybe we can look at other possibilities."
- "I guess in some ways I understand you. We need to talk some more about specifics."
- "Sometimes you feel like I don't really understand what you are saying. Is that it?"

Exercise 8-3
PULLING IN THE CLIENT-COUNSELOR RELATIONSHIP

Three participants are needed for this exercise—a client, a counselor, and a doubling counselor. The client presents a problem, either one of his own or one of the situations given below.

The client is free to discuss his problem in any way he wishes. The counselor, however, may answer in only one sentence. The doubling counselor listens to both the client and the counselor. If the doubling counselor does not think that the counselor has accurately dealt with the client's statement, he may add to what the counselor has said, but with only one sentence. Anyone in the rest of the group who is still unsatisfied with the responses to the client may tap the doubling counselor on the shoulder and he leaves. Then the new doubling counselor adds what he considers appropriate. This counseling session should take no more than 10 minutes. Then the third column (immediacy) of Table 5 is used to determine the functioning level of the counselor and the doubling counselor. After the discussion a second counseling session of no more than 10 minutes' duration is conducted and the evaluation procedure is repeated.

1

Two weeks ago Helen gave birth to a boy. She is disturbed by her ambivalent feelings of love and hate for the infant. When she sees him, she wants to touch him and hold him, but then she thinks of how annoying his crying is and she "can't stand him." She is afraid she might hurt the baby if she touches him because her "hatred is so bad." She phones a counselor to talk to him about what she should do.

2

Jan is a weekend social drinker. Although he spends most of his time drinking over the weekend, he doesn't consider himself to be an alcoholic, but "just one of the boys." "After all," he comments to his wife, "I do have a responsible position and only drink a little with my business associates." He refuses to see anyone about his drinking. "Don't need to," he says. His wife, Martha, is upset and is seeking help from Dr. M.

Exercise 8-4
GETTING DEEPER

The group divides into pairs. Each pair role-plays all of the following situations. Afterwards, the fourth column (self-exploration) in Table 5 is used to determine the functioning level of the counseling situation.

1

Two people talk to each other with nonsense syllables (mft, hpt, and so on), not words. Each person tries to convey his feelings as well as his ideas with his nonwords for 5 minutes. Afterwards the two participants discuss for 5 minutes whether the nonword messages were understood accurately.

2

Person A is from Earth and person B is from Mars. Person A must interview person B for 5 minutes for an article to appear in a newspaper. He must find out how person B feels about meeting A, an earthling. Afterwards, persons A and B discuss for 5 minutes how effectively person A conducted the interview and whether he got information from person B about his feelings.

3

The client tries to conceal some thought and/or feelings from the counselor. For 10 minutes the counselor encourages the client to open up and explore this feeling or idea.

• • •

The words fuse into each other when considering feelings, concreteness, immediacy, and self-exploration. These have been divided into separate categories for discussion purposes, yet these words are inseparable and interdependent. Empathy is the catch-all word for all of the characteristics of accurate understanding.

Genuineness

Two words are closely related in explaining the meaning of genuineness—congruence (when one's words and actions correspond) and authenticity (when one is himself, not a phony). Basic to all three of these words is the idea of being real. The most *real* individual is the infant who has not as yet learned the mask of politeness or the unreliability of distrust. The infant is born *genuine* (to be natural) and is conditioned to adopt a front that diminishes self-disclosure. In Fig. 8-1, Maslow's (1954) and Erikson's (1950) concepts of the stages of growth toward self-fulfillment are diagrammed. Both Maslow and Erikson stress that an individual must turn on to himself before he can turn outward toward others. As the steps of the pyramid in Fig. 8-1 show, there is a gradual development of the child from self-interest to interest in

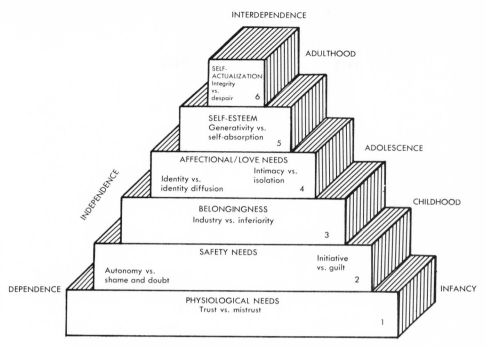

INTERDEPENDENCE

ADULTHOOD

SELF-
ACTUALIZATION
Integrity
vs.
despair   6

SELF-ESTEEM
Generativity vs.
self-absorption   5

ADOLESCENCE

AFFECTIONAL/LOVE NEEDS

Identity vs.
identity diffusion

Intimacy vs.
isolation   4

CHILDHOOD

BELONGINGNESS
Industry vs. inferiority   3

INDEPENDENCE

SAFETY NEEDS

Autonomy vs.
shame and doubt

Initiative
vs. guilt   2

DEPENDENCE

PHYSIOLOGICAL NEEDS
Trust vs. mistrust   1

INFANCY

Fig. 8-1. Stages of growth toward self/other fulfillment.

others. This *natural* characteristic at birth is distorted if:

- The need for air, food, water, physical comfort, rest, and activity is not satisfied
- The need for freedom from fear, insecurity, and danger and threat are not accomplished
- The need to be *somebody*, first in the family group and later on in other groups, is not experienced
- The need to be loved (cared for and comforted) and to love is not encountered
- The need to feel worthwhile and competent is not achieved

Out of the gradual satisfaction of these needs emerges a feeling of fulfillment, an outward seeking to know—to understand oneself and later to understand the world and other people. All of the needs are present at each stage of personal growth. However, certain needs are more important at certain ages. The redevelopment of genuineness therefore requires that the individual learn to trust, to feel autonomous (self-directed), to be comfortable in trying out new experiences (initiative), to work at learning and achieving (industry), to be developing who "I am" and becoming "who and what I

want to be" (identity), and to care for oneself and for others (generativity). In essence, to be genuine is to be more oneself.

Definition

Genuineness is the process of self-disclosure in which an individual is sufficiently aware of and accepts his own feelings and experiences so that he is able to freely and appropriately communicate his feelings and experiences to someone else; he is also open to the communication of feelings and experiences from the other person. In addition to openness to oneself and to others, the genuine person develops the ability to honestly tell the difference between his own feelings about the experiences of the other person and the other person's experiences of his own feelings. Thus, when the genuine person gets caught up in thinking, "What a jerk she is. She thinks she's superior to me," he knows he is surrounded by his own prejudices. He must sift through his feelings and see the other person's point of view: "She does put on a superior air. I wonder whether she feels comfortable with me. Something about the way she looks at me makes me wonder."

General characteristics

Out of genuineness arises self-understanding, understanding of others, and a willingness to be known. The definition of genuineness suggests several other general characteristics.

Correspondence. Defensive feelings are often barriers to free and open communication. For example:

1

Ann fears rejection from Hank.

HER FEELINGS: "I love him, really love him. But, does he love me?"

HER STATEMENTS: "Come on now, let's not get too close into things. Won't work."

HER ACTIONS: She bites her lower lip, tightens her jaw, and sits back in her chair with her legs loosely placed on the floor, one ankle bent, the other leg stretched out.

2

Hank doesn't want to get too involved with anyone right now, but he would like a casual sexual arrangement with Ann.

HIS FEELINGS: "Better watch out for this chick. She's scrounging for ties."

HIS STATEMENTS: "Why not? Getting close is what it's all about, pleasant being, got a lot to share."

HIS ACTIONS: He pulls his lip up to one side in a half smile, raises one eyebrow slightly, squints his eyes and leans forward with his hands tightly clasped between his knees.

• • •

Defensive maneuvers are evident in both Ann's and Hank's statements. The words spoken evaluate the relationship instead of describing how Ann and Hank feel. Ann says that getting close won't work. She is implying that she fears taking a chance on Hank's uncertainty. Apparently Ann would like to control the interrelationship in order to defend herself against rejection and emotional hurts. She reveals by her facial and body cues that she is not leveling with Hank. She tenses her face, stretches out one leg, and partially pulls the other leg back at the ankle. Instead of speaking her feeling, she says with certainty what she is afraid might happen ("won't work").

Hank is also defensive and not leveling with Ann. He tries to cover up his own concerns with some tactics that guard against revealing his flight from real intimacy and self-disclosure. He pretends he wants closeness, yet he is pulling himself both toward (leans forward in chair) and away from Ann (hands tightly clasped between his legs).

Ann and Hank are game-playing. Game-playing is the artificial social device that individuals use to relate to one another in order to avoid being hurt. Game-playing means you either "don't make waves" or when you do, you arrange the waves carefully. Game-playing for the counselor may mean that he does not let his humanness show; he plays the counseling role with only his techniques showing. Under these conditions the counselor cannot say what he means nor can he mean what he says. The counselor, in other words, is not genuine. Exercise 8-5 directs attention to the need for correspondence between verbal and nonverbal behavior.

Exercise 8-5
MAKING VERBAL AND NONVERBAL BEHAVIOR GO TOGETHER

This exercise starts from what should not be and then goes into what should be genuineness. In this way, by going from "wrong" to "right," an individual becomes more aware of how phony behavior interferes with the counseling transaction.

Arrange a group of five participants. Two of the participants are to be a client and a counselor. The three other participants are observers. One individual observes nonverbal behavior, one individual observes verbal behavior, and one individual observes the correspondence between verbal and nonverbal behavior. The observers need two copies of the rating scale (Table 6) in order to rate the counselor on congruence.

The counselor selects one of the situations below and the client role-plays the problem. The counselor exaggerates his behavior to show his self-interested purpose—to do his *worst* in order to get a rating at the lower end of the interpersonal scale. The counselor keeps before him the idea that he hates this role-playing requirement and is really not interested in the client. In fact, he thinks the person role-playing the client is a jerk. The counselor must be as phony as he can be.

After role-playing, the group discusses for 10 minutes the ratings of the counselor's correspondence and points out the verbal and nonverbal behavior that reveal the counselor's lack of interest in and respect for the client. Examples of the lack of correspondence between verbal and nonverbal behavior also are discussed. Group members make suggestions for the counselor's improvement. Then the same incident is role-played for 5 more minutes. This time the counselor respects the client, is interested in him, and thus is genuine and trustworthy. The three other participants shift their observations so that they are observing different behavior than during the last counseling episode—for example, a participant might observe verbal behavior this time if he observed nonverbal behavior last time. The observers rate the counselor on this second role-playing; a second general discussion follows in which the observers include examples of the counselor's improvement in genuineness (correspondence between verbal and nonverbal cues).

1

Mr. P. is counseling a 30-year-old Mexican-American man who has been having difficulty in getting a job. Mr. P. does not believe that boycotting lettuce or grapes is a smart thing to do. "That union guy Chavez causes so much unnecessary trouble." Mr. P. thinks as he looks at the brown skin and blue shirt and jeans of the client, "Mess, what a mess. That's the trouble with these Chicano guys. Talk about machismo, macho [manliness], and are really just weak sisters. If a man wants to work and better himself, he can get a job, an education. This union stuff is un-American. Gotta be nice to the guy; damn supervisor is all for this minority stuff."

2

Ms. I. believes she is a liberated woman who is "unshockable." However, she really does get disgusted with these drug freaks who try everything. "These young longhairs who make the drug rounds like this 15-year-old, flipped-out kid, Marcia, seeing me, can't take the world. Got to make her see I really understand and accept her." Marcia is telling Ms. I. that she has

been smoking "angel dust"* and feels ". . . far away, moving back and forth from my arms and legs; can't find my legs. Mmmmm, feel, mmmmmmm, so empty—shrinking, shrinking. Oh my God, I won't beeeeeeee . . . ! Got to find my shoes, my feet." Ms. I. is displeased and thinks, "Don't want to handle another one of those. But who else can do it?" She says, "I'm here with you, Marcia."

Self-disclosure. Correspondence between verbal and nonverbal behavior is essential to genuineness, and self-disclosure is a natural outcome of this correspondence. Self-disclosure signifies the willingness to be known.

Self-disclosure may be simply defined as the offer of personal information. In the counseling situation this offer must be according to the client's desire to know about the counselor as well as the counselor's understanding of which of his experiences would best fit the client's needs. The counselor's willingness to share himself with the client encourages similar self-disclosure by the client. Mutual exchange of information is encouraged.

In general, people need to feel free to reveal information as well as to conceal; this applies to the counselor as well as to the client. The main difference between the degree of concealment is that the counselor, as compared with the client, has more open space (see discussion of Johari's window on p. 54) that he can comfortably share with the client *for the client's benefit*. This last phrase is very important, for the purpose of the counselor's disclosure is to prompt similar disclosure by the client. The goal is not a confession to absolve the sins of either the

---

*Angel dust is the name sometimes given to phencyclidine (PCP), a psychoactive or consciousness-expanding drug that depresses the central nervous system and results in a state resembling alcohol intoxication, with muscular incoordination, generalized numbness of the extremities, sensory disturbance, muscular rigidity, and probable loss of contact with one's environment. Large doses may produce convulsions. Sometimes other side effects such as vomiting, nausea, drowsiness, confusion, and other nonpleasant effects result. PCP is also referred to as "PeaCe Pill" and Hog (Stash notes, 1973).

counselor or the client. The goal is mutual sharing for the client's benefit.

Confrontation. From the counselor's standpoint, self-disclosure serves as a form of self-confrontation. Not only does the counselor reveal to the client his experiences but also, in this process of revealing, the counselor must consider and weigh the meaning of his own experiences. Facilitative (effective and reinforcing) counselors are more alert to and inclined to tell about differences in their emotional experiences. More contented and more satisfied counselors, in fact, are more able to identify, accept, and respond nondefensively to their own reflections and disclosures.

Confrontation has at least two purposes. It provides informational feedback for both the client and the counselor. At first, feedback (confrontation) arises from the counselor's efforts to discover himself; then it comes from the counselor to the client. Later the client begins to give feedback to the counselor, and finally both the counselor and the client give feedback on the feedback.

Confrontation was discussed in Chapter 7 as one of the expressive skills. The explanation, general purposes, influential variables, and procedures involved are applicable to the present discussion of confrontation. Example 8-1 describes the behavior of an alcoholic who exhibits the usual low level of self-confrontation, which is supported by his alcoholic reasoning ("alcologia").

Example 8-1
CONFRONTING "ALCOLOGIA"

Mr. R. is certain that his wife is wrong. He is not an alcoholic. He thinks to himself, "Why, just the other day, last Friday, it was, when I realized that all that beer drinking was just making me gain too much weight, when I was getting so fuzzy. Couldn't work. So what did I do? I cut down on liquids; 'No more beer,' I said, 'Take too much.' So decided to drink bourbon and soda. Don't drink as much now. My wife, she just doesn't understand the willpower I have."

For a few months he drinks bourbon and soda only as a pick-me-up when he gets up in the morning and a few to be social with the fellows at lunchtime, and, of course, just a couple before dinner. When his wife

"nags him" because he "doesn't look well," he has to "drink away" his "uncomfortable feeling." He decides to switch to bourbon and water. "After all, it must be the carbonation in the soda. That fizz makes me have indigestion."

His wife tells him she can't take his behavior any more and that she has gone to some meetings of Al-Anon. Maybe he ought to go to that doctor who treats alcoholics. That will shut her up, at least. After all, all he needs to do is stop drinking the bourbon with the water. It is the water that is giving him the gas, his embarrassing hiccups, and resounding burps. "So, I'll drink bourbon on the rocks. Who says you have to drink bourbon with soda or water?" considers Mr. R.

At first, Ms. R. pours all the bourbon down the sink so that Mr. R. will not find any to drink. She soon discovers that this is foolish because Mr. R. either manages to have a bottle hidden in the oddest places, such as the baby's diaper pail or behind the toilet bowl, or he travels miles, if necessary, to get some. She also is comforted by the Al-Anon meetings, which assure her that she is not at fault. It is Mr. R.'s responsibility. All she can do is to be there when he needs her and show that she cares for his welfare. "The situation might go on forever," Ms. R. sadly supposes, "or at least until his liver deteriorates completely." However, Mr. R. begins drinking more to "feel comfortable" or to "tide himself over the ugly feeling between drinks" and goes home later. One evening he remains at the bar a little longer than usual, drinking even more than usual. In fact, he is now drinking bourbon straight since when the "rocks" melt he is still getting too much liquid.

Mr. R. lives in the suburbs and in the 10 years that he has been working and drinking, he has managed to escape having any serious accidents. Tonight he wobbles from the bar to his car in the garage near his office. "Sure, I'm fuzzy, must be tired," he slurs. He pulls out of the garage and slams into an oncoming car.

In the hospital, Mr. R. awakens to see his wife and some bearded guy. Despite the pain in his bandaged head, Mr. P. manages a question, "Who're you?" The bearded man answers, "I'm Dr. McNeil; your wife has asked me to see you. Just saying hello today. I'll be back tomorrow when you are feeling a little better.''

Exercise 8-6
CONFRONTING "ALCOLOGIA"

"Alcologia" is the term used "to describe the combination of denial and rationalization typical of alcohol reasoning" (Twerski, 1973). The importance of confronting Mr. R. with his inconsistencies is crucial. It may be difficult to make a wedge in the distorted thinking that is part of "alcologia," yet it is essential to do so in order for the alcoholic to learn to face his responsibility for the act of drinking.

Two trainees role-play this exercise. One trainee role-plays Mr. R. and the other trainee role-plays Dr. McNeil. The rest of the group should each have a copy of the scale for rating functioning level of genuineness (Table 6) and two copies of the score

sheet for genuineness. The observers take notes on the counselor's confrontation incidents and note how the client accepts or rejects the counselor's confrontations. The counseling session may take up to 30 minutes. After the group rates and scores Dr. McNeil, a general discussion follows of Dr. McNeil's confrontation procedures and his level of genuineness. Suggestions are made for improvement.

The entire procedure is then repeated for 15 minutes. The discussion after the rating of Dr. McNeil's genuineness should include examples of how he has improved in confrontation procedures and in level of genuineness.

Genuineness scale (Table 6)

Specific characteristics. The scale for measuring the functioning level of genuineness incorporates the five progressive levels

Table 6. Scale for rating functioning level of genuineness*

CONGRUENCE (CORRESPONDENCE)	SELF-DISCLOSURE	CONFRONTATION
LEVEL 1		
C.'s verbal responses are clearly defensive and unrelated to what he is feeling at the moment, as shown in what he is saying and in the sound of his voice. Contradictions appear in what he says. C.'s only genuine responses are his negative (unfavorable) remarks to the client.	C. remains removed from Cl. and is close-mouthed about his own feelings or personal beliefs. If C. does disclose himself, he does so in a way that is not tuned in to Cl.'s needs and may make disclosures about himself that may disturb Cl. C. tries to turn off Cl. from asking personal questions.	C.'s verbal and nonverbal behavior disregards or passively accepts inconsistencies in Cl.'s behavior.
LEVEL 2		
C.'s verbal responses are slightly unrelated to what he is feeling at the moment. When C. does respond appropriately with the right words, his voice and general manner have a "canned quality," as if he were reading from a book or imitating a "professional air." Thus C. responds carefully and correctly but without the expression of real, felt emotions. C. may respond genuinely but with negative remarks concerning	Although C. does not always appear to be actively avoiding self-disclosure, he never offers any personal information about himself. C. may respond briefly to direct questions from Cl. but does so unwillingly and answers only what Cl. specifically requests.	C. shows by his verbal and nonverbal behavior his disregard of Cl.'s inconsistencies in behavior. C. remains silent about Cl.'s inconsistencies yet shows by some behavioral expressions that he is aware of and not accepting the inconsistencies. However, he does no more than hint at his awareness.

Continued.

Table 6. Scale for rating functioning level of genuineness—cont'd

CONGRUENCE (CORRESPONDENCE)	SELF-DISCLOSURE	CONFRONTATION
LEVEL 2—cont'd		
Cl. C. does not seem to know how to use his negative reactions to Cl. as a basis for exploring the relationship between C. and Cl.		
LEVEL 3 (minimum facilitative level)		
Although C. does not show any real inconsistency between what he says, how he says it, and what he does while saying it, he still does not give any definite cues that his response is genuine. C.'s behavior hints at possible underlying defensiveness or professionalitis (Chapter 6).	C. offers personal information about himself, which may be tuned in to Cl.'s needs. However, this information is often indistinct and too general to really describe C. C. does give impression that he is willing to disclose more about himself but is less free in dealing with his feelings about Cl. and about the transactions going on between them.	C.'s verbal and nonverbal behavior reveals that he is aware of the inconsistencies in Cl.'s behavior, but C. does not relate directly and specifically to these inconsistencies. C. may ask questions about inconsistencies without pointing out the disagreements and/or inconsistencies in Cl.'s answer.
LEVEL 4		
C. responds with many of his own feelings and is positively genuine and nondestructive in his responses to Cl. What C. says corresponds with what he is feeling, but he is somewhat hesitant to express his feelings fully. C. is able to handle his favorable as well as his unfavorable feelings toward Cl. as a basis for further exploration of the interpersonal relationship. There is no evidence of false front, of defensiveness, or of professionalitis by C.	C. freely offers personal information about his ideas, attitudes, feelings, and experiences that fit into Cl.'s interests and concerns. In fact, C. may discuss intimate ideas in both depth and detail, and his expressions clearly reveal him as a distinct individual.	C.'s verbal and nonverbal behavior attend directly and specifically to inconsistencies in Cl.'s behavior. When C. confronts Cl., he does so honestly and sensitively.
LEVEL 5		
C. is freely and deeply himself in an unselfish, caring relationship with Cl. C. is nondefensively open to experiences of all types, both pleasant and hurtful, and is able to use these experiences comfortably and constructively for deeper exploration for himself and for Cl. Verbal and nonverbal expressions match.	C. offers even intimate and detailed information about himself in tune with Cl.'s needs. C. gives impression that he is holding nothing back and freely presents both his ideas and feelings to Cl.	C.'s verbal and nonverbal behavior is sharply tuned in to the inconsistencies in Cl.'s behavior. C. confronts Cl. in an honest, sensitive, understanding manner when inconsistencies in Cl.'s behavior appear.

*C. = counselor; Cl. = client.

SCORE SHEET FOR GENUINENESS SCALE

Place check under *one* of the levels for *each* characteristic. The genuineness level equals the total number of checks under each category divided by 3, the total number of categories. This is the average level of all the characteristics.

EXAMPLE:	Congruence	Level 1
	Self-disclosure	Level 2
	Confrontation	Level 2
	Total =	5
	Divided by 3 = 1.7 (level 1+)	

Characteristics	Levels				
	1	**2**	**3***	**4**	**5**
Congruence (correspondence)					
Self-disclosure					
Confrontation					
Total No. of checks under each category					

*Minimum facilitative level.

of congruence, self-disclosure, and confrontation (Truax and Carkhuff, 1967). The interrelationship among these three characteristics is evident when one looks at one of the levels described. At level 1, if the counselor is clearly defensive and contradictory in his responses to the client (low level of congruence), he is more likely to be close-mouthed in voicing his feelings and beliefs (low level of self-disclosure) and also less likely to be aware or, at any rate, less likely to reveal his awareness of the client's inconsistencies in behavior. On the other hand, the counselor may point out the inconsistencies in a manner destructive to the client's self-esteem. The counselor's own inconsistencies are covered up and the client's inconsistencies are ignored or harmfully reported.

Nonpossessive warmth

Empathy, genuineness, and nonpossessive warmth are the triad of interpersonal skills that influence the client's personal growth favorably or unfavorably. It is doubtful whether excellence of skill in any one of the characteristics involved in the Truax triad would be sufficient for effective counseling. Confrontation alone will not lead to increased self-exploration (Kaul et al., 1973). Respect for the client, a general characteristic of warmth, is also unlikely to be the sole

determinant in furthering the client's eagerness and ability to accomplish more successful coping behavior.

There is an interdependence and interchange of effects between all the characteristics of the Truax triad. Perhaps greater development of any one skill may increase the possibilities for a more successful client outcome. However, an excess of any one characteristic may actually be destructive because it may temporarily overshadow actual counselor deficiencies. These points are further discussed later in this chapter.

Definition

There are two key words that require definition—"nonpossessive" and "warmth." The term "nonpossessive" may be more easily defined if the "non" part of the word is temporarily removed and "possessive" is examined first. "Possessive" brings to attention both ownership and the occult or mystic meaning of possession, that is, control by an invading being or thing. These two ideas together give a more complete meaning. Both the materialistic notion of using an individual for one's own gain or profit (ownership) and the manipulative notion of control by someone other than oneself are combined. Nonpossessive is obviously the opposite of these notions.

Nonpossessiveness is unconditional, that

is, with no strings attached. Nonpossessiveness refers to self-direction, self-control, and consequently self-choice in decision making. Nonpossessiveness requires that the counselor be open to the client's point of view and that he avoid imposing upon or taking over with his own viewpoint.

Nonpossessiveness puts limitations on warmth only insofar as the feeling of warmth must be expressed for the client's welfare, not for the glory of the counselor. Warmth goes beyond involvement; it takes in the tender feeling of caring and the comfortable freedom of expressing feelings with no limits. The counselor who expresses warmth is acceptant. He accepts the client "where he's at" at the moment of the counseling session.

Nonpossessive warmth, accordingly, may be defined as the expression of commitment, concern, and respect for another human being *as he is, not as he ought to be*. Nonpossessive warmth is a positive feeling of closeness that overlaps and intertwines with accurate empathy and genuineness.

General characteristics

Nonpossessive warmth is more difficult to explain than empathy and genuineness since it is so dependent on the expression of feelings. Transactional analysis explains warmth as positive stroking, which is preferably unconditional. Learning theorists stress warmth as reinforcement of appropriate behavior by means of some form of approval. Several characteristics are included in the concept of warmth—equal worth, absence of blame, nondefensiveness, reciprocal affect, and closeness. The list may be extended even further, but these seem to be the characteristics mentioned most often in the literature. Other characteristics are further extensions of those just mentioned. In fact, the listed items overlap and are divided primarily for discussion purposes.

Equal worth. Equal worth is basic to the quality of warmth, and acceptance is the related factor. Acceptance embraces human equality as one of its elements. If one individual considers another individual equally worthy, it follows that he will uphold the other person's right to be different. This acceptance of difference is not attached to approval or agreement with the individual's difference. Nevertheless, no weight of lesser or greater worthiness is given to the other individual's self-experiences. For the counselor, viewing the client as having equal worth promotes a positive regard for the client, and this regard does not fluctuate with the counselor's measurement of the value of the client's life-style.

Nonblame. Equal worth demands a nonjudgmental attitude, and this nonjudgmental attitude results in absence of blame. Nonblame, however, does not suggest the client's lack of responsibility for his acts. Instead, nonblame suggests that if the counselor were in the client's circumstances he would be likely to act in a similar fashion. Nonblame may also imply that the client is accepted even though he feels unacceptable. Underlying this concept of nonblame is the idea that the counselor accepts the client because *he is*.

Therefore the client feels safe enough to freely express himself in the nonthreatening counseling situation. In an atmosphere of safety from blame the client learns to trust the counselor and also himself. He moves toward greater assurance in his ability to handle his own destiny.

Nondefensiveness. Through self-understanding the counselor becomes more open to the client. If the counselor attains sufficient insight about his own needs and how he may fulfill these needs, he is less likely to expect the client to satisfy his needs. The counselor is also more likely to be open to the client's background, goals, and value system if he does not feel defensive about his own background and beliefs.

Another outcome of the counselor's self-understanding and self-acceptance is his increased ability to honestly reveal his feelings to the client. The counselor is able to air his feelings of pleasure as well as those of anger. Properly balanced doses of warmth and anger may actually induce greater cooperation from the client (Johnson, 1971a), particularly if the counselor's angry response is immediately followed by assurance of his warmth.

Revealing anger clears the air and admits the counselor's humanness. However, the expression of anger must be accomplished with caution. The empathic counselor is sensitive to the impact on the client of the counselor's disclosure of feelings. Anger or any other emotional expression may alienate the client (turn the client away from the counselor) unless this expression is handled sensitively.

Reciprocal affect. Another factor that the counselor must consider is referred to as the principle of reciprocal affect (Truax and Carkhuff, 1967). This principle states that when one person expresses a certain emotion (affect), this emotion is likely to call forth a similar emotion in the other person (reciprocal). Accordingly, anger encourages anger, joy encourages joy, sorrow encourages sorrow, defensiveness encourages defensiveness, and so on. Unless one of the individuals alters this snowballing effect of reciprocal affects, the emotional intensity builds up. In the counseling situation it is the counselor who must cut the anger short by means of a change in his own behavior. For instance:

EVA: It's that 18-year-old clod who is causing the trouble. We'd be okay. Lazy, good-for-nothing goof-off. Won't work, won't go to school, talks only of people power—demonstrating, noisemaking. What does he want? He's got a beautiful home, so he doesn't like the suburbs, stifles him, middle-class morality. Lives with a tramp; she's 26 years old and has a child. Both of them I spit on.

MS. G.: Wow! You've really got it in for him. Have you ever considered his side of the picture? His values?

EVA: Whaddya know about a son? You're too young to know what it's like; you. . . . Well, you don't know what you're saying.

MS. G.: Hell! You have a lot of nerve saying that to me. I have studied to work with people. You don't know any. . . . Heavens, what am I doing? I really am angry, really lost my cool. Let's look at this again.

The counselor who is nondefensive is able to admit his errors to the client. "I really jumped to conclusions just then" or "There I go, giving my opinion instead of listening to yours."

When the counselor reacts to the intense feelings of the client, as Ms. G. does, the counselor's perceptions (ways of seeing, hearing, and feeling) of the client become blocked. Not only is the counselor less able to listen to the client but he is also less able to recall what the client has just done or said. Too much threat increases the anxiety level of both the client and the counselor to the point where they may both lose sight of the purpose of the counseling session.

Closeness. Anger and/or anxiety may have another effect that detracts from warmth. These negative emotions tend to make the psychological distance greater between counselor and client. A more distant (standoffish) relationship between counselor and client turns into coldness and less give and take between counselor and client.

Closeness has two possible explanations, depending on the direction the closeness takes. When an individual is close to himself, he leans toward a tight-lipped secretiveness and keeps to himself and away from others. Such a person is apt to maintain an area of psychological and probably also physical space around him. In other words, he is "cold" and keeps his distance from others.

The opposite meaning for closeness shrinks the person-to-person distance. Closeness implies coming in contact either psychologically, through understanding and acceptance, or physically, through touch. Often both physical and psychological contacts coexist. This meaning of closeness suggests a fellowship of warmth. When one person says of another, "He is someone with whom I dare to be me" or "She knows my faults and yet she likes me," then closeness exists. It is this sense of the word "closeness" that is essential to the teammate companionship of the effective client-counselor relationship. Furthermore, it is easier to connect, to rap, with someone who considers one worthy of respect, who does not blame, who really tries to understand, who is not defensive about what he says, and who does not closet himself into his own thinking but instead comes close enough to share.

There is one note of caution about warmth. The counselor who feels and exhibits warmth increases his personal attraction to the client and consequently summons up greater cooperation from the client. However, an overabundance of warmth associated with excessive liking may lead to an overdependent client. If this should occur, the counselor should examine his motives—is he actually striving for a dependent client?

Exercise 8-7 is arranged so that an individual may examine the degree of liking and warmth he feels toward different people and how these same people feel about him. The participant should keep in mind that this is not a scientifically designed instrument and the score merely offers a basis for comparison and discussion.

Exercise 8-7
RELATING LIKING AND WARMTH

This exercise may be accomplished by either procedure A or procedure B.

Procedure A

Select at least two people each day for a minimum of 3 days (a total of six people). These people must be willing to rate and score how *they* feel you respond to them in addition to your scoring of how *you* feel they respond to you. Each day you should select someone you like and someone you dislike or, at any rate, do not like very much. Do not discuss your feelings of like or dislike for these individuals before you and they have completed the ratings and scoring. Such a discussion may influence you and may alter their responses toward you. Afterward, you may do as you think best about discussing your relationships with the individuals you have selected.

Procedure B

Select six people in your training group, three whom you like and three whom you dislike. All seven of you use the rating and scoring forms to determine the warmth ratings for your relationship.

Ratings

In each situation, when you do the rating or when the other person rates, these are to be in terms of self-rating. Thus you are to rate the other person as you feel he responds to you, and the other person is to rate you as he feels you respond to him. For instance, Tom asks Mary to rate him. Tom rates Mary's feelings for him, and Mary rates Tom's feeling for her.

Cares ___:_x_:___:___:___ Doesn't care (Tom rates Mary.)

Cares ___:___:_x_:___:__ Doesn't care (Mary rates Tom.)

RATING SHEET FOR WARMTH

List of word opposites
Remember. These words refer to the other person's feelings about you.

1. Cares	:___:___:___:___:___:	Doesn't care
2. Likes	:___:___:___:___:___:	Dislikes
3. Appreciates	:___:___:___:___:___:	Doesn't appreciate
4. Friendly	:___:___:___:___:___:	Unfriendly
5. Interested	:___:___:___:___:___:	Uninterested
6. Values me	:___:___:___:___:___:	Doesn't value me
7. Patient	:___:___:___:___:___:	Impatient
8. Accepts	:___:___:___:___:___:	Rejects
9. Close	:___:___:___:___:___:	Distant
10. Shares thoughts	:___:___:___:___:___:	Doesn't share thoughts
11. Shares feelings	:___:___:___:___:___:	Doesn't share feelings
12. Shares himself	:___:___:___:___:___:	Doesn't share himself
13. Acts personal	:___:___:___:___:___:	Acts impersonal
14. Nonjudgmental (doesn't blame)	:___:___:___:___:___:	Judgmental (blames)
15. Nondefensive	:___:___:___:___:___:	Defensive
16. Open	:___:___:___:___:___:	Holds back
17. Honest	:___:___:___:___:___:	Dishonest
18. Concentrates	:___:___:___:___:___:	Doesn't concentrate
19. Kind	:___:___:___:___:___:	Unkind
20. Calm	:___:___:___:___:___:	Angry

In other words, Tom feels that Mary cares for him some of the time so he places an **x** in the second box. Mary is uncertain of whether or not Tom cares for her so she places her x in the third or middle box.

Most of the time: Place __x__:___:___:___:___ in first box.

Some of the time: Place ___:__x__:___:___:___ in second box.

Uncertain: Place ___:___:__x__:___:___ in middle (third) box.

Some of the time doesn't care: Place ___:___: ___:__x__:___ in fourth box.

Most of the time doesn't care: Place ___:___: ___:___:__x__ in fifth box.

Use the list of words in the rating sheet on p. 182 for rating the individuals you have selected for this exercise. After you have finished rating the other person's responses to you according to the 20 word responses listed, record the results on the score sheet below. To obtain the total warmth score, add up the numbers in the plus column of scores and place the total in the box at the bottom of the plus list. Disregard the minus sign for the moment and add up the numbers in the minus column and place the

SCORE SHEET FOR WARMTH RATING

Key: +2 = most of the time favorable   −1 = some of the time unfavorable
      +1 = some of the time favorable   −2 = most of the time unfavorable
                        0 = uncertain

Word opposites	+2, +1	Scores 0	−1, −2
1. Cares—doesn't care			
2. Likes—dislikes			
3. Appreciates—doesn't appreciate			
4. Friendly—unfriendly			
5. Interested—uninterested			
6. Values me—doesn't value me			
7. Patient—impatient			
8. Accepts—rejects			
9. Close—distant			
10. Shares—doesn't share thoughts			
11. Shares—doesn't share feelings			
12. Shares—doesn't share himself			
13. Acts personal—acts impersonal			
14. Nonjudgmental—judgmental			
15. Nondefensive—defensive			
16. Open—holds back			
17. Honest—dishonest			
18. Concentrates—doesn't concentrate			
19. Kind—unkind			
20. Calm—angry			
Totals	+	0	−

WORDS	EX. 1: PLUS SCORE			EX. 2: ZERO SCORE			EX. 3: MINUS SCORE		
	+	0	−	+	0	−	+	0	−
1	+2			+2					−2
2	+2			+2					−2
3	+1				0			0	
4	+2					−1		0	
5	+1					−2	+2		
6		0				−2			−2
7			−1	+1				0	
8	+1					−2			−2
9	+2			+2				0	
10			−1	+2					−2
11			−1		0				−2
12		0			0				−1
13	+1					−2			−2
14		0				−1	+2		
15	+1					−1			−2
16	+1				0				−2
17	+1			+2					−2
18	+2				0				−2
19			−2			−1	+2		
20			−1	+1				0	
Totals	+17	0	−6	+12	0	−12	+6	0	−23

TOTAL WARMTH SCORES: Ex. 1. +17− 6 = +9
Ex. 2. +12−12 = 0
Ex. 3. + 6−23 = −17

total in the box at the bottom of the minus list. Subtract the lower total (plus or minus total) from the larger total. This is your warmth score. Your warmth score may be either positive (+), negative (—), or even occasionally zero (0). See the examples above. In order to compare your own scores with the score of the other individuals, write both sets of scores on the chart on p. 185.

Discuss the results, answering the following questions.

1. Do you feel that the list of words actually represents liking and/or warmth as you perceive these concepts? If you believe that certain items are better than others, point these out. If you think certain items should be removed and others inserted, point this out.
2. Is there a difference in the warmth ratings for the people you like and for those you do not like?
3. Is there a difference in the warmth ratings for you from the people you like and those whom you do not like?
4. What has this exercise told you about yourself? What has this exercise told you about the relationship between your liking and warmth to the responses of other people?

	MY OWN SCORES			OTHER PERSON'S SCORES		
	+	0	−	+	0	−
People I like A						
B						
C						
People I don't like D						
E						
F						

Table 7. Scale for rating functional level of warmth*

POSITIVE REGARD	RESPECT
LEVEL 1	
C. actively offers Cl. advice and is indifferent to and uninterested in Cl. C. is heedless of Cl. as a person. C. feels responsible for Cl. and often indicates that he knows what would be best for Cl. C. is actively critical of Cl., approving or disapproving of Cl.'s behavior. C. may in fact dislike Cl. and reveal his impatience with Cl. C.'s over-concern for Cl. interferes with C.'s open and clear discussion and with Cl.'s free responses.	C.'s verbal and nonverbal expressions communicate a clear lack of appreciation and esteem for Cl. C. conveys to Cl. that Cl.'s feelings and experiences are not worthy of consideration and that Cl. is incapable of acting constructively. C. is more concerned with himself and shows his contempt of Cl. In fact, C. may focus on himself to such an extent that he tells Cl. more about his own opinions and feelings than Cl. wants to know. Cl. is actually concentrating on "blowing his own horn," on increasing his own self-respect.
LEVEL 2	
C. ignores Cl., showing little interest or genuine kindness. C. responds mechanically and is often passive, as if he were not attending to Cl. Sometimes C. is cold and disapproving, but C.'s feelings toward Cl. vary considerably and are dependent on Cl.'s response to him (C.).	C.'s verbal and nonverbal expressions indicate lack of appreciation and esteem for Cl. However, C.'s interest in and response to Cl. depends partly on what Cl. is talking about and how Cl. is feeling about himself; at times, C. may respond with some recognition and appreciation of Cl.'s worth and at other times C. may be unresponsive to Cl.'s worth as a person.
LEVEL 3 (minimum facilitative level)	
C. shows positive caring in that he communicates to Cl. that his (Cl.'s) feelings and behavior matter to him (C.). C. sees himself as responsible for Cl. and is thus semi-possessive, telling Cl., "I want you to . . ." and "It's important to me that you. . . ."	C.'s verbal and nonverbal expressions indicate some degree of appreciation and esteem for Cl.'s feelings, experiences, and potentials. C. shows that he values most, if not all, of Cl.'s opinions and expressions about himself.
LEVEL 4	
C. shows deep commitment, interest, and concern for Cl.'s welfare and accepts Cl. as a person free to be himself with little evalua-	C.'s verbal and nonverbal expressions indicate appreciation and esteem for Cl.'s feelings, experiences, and potentials. C.'s responses

Continued.

Table 7. Scale for rating functional level of warmth—cont'd

POSITIVE REGARD	RESPECT
LEVEL 4—cont'd tion or criticism of Cl.'s beliefs or feelings. However, C. conveys to Cl. that he expects Cl. to be mature (adult) and not to regress (retreat into earlier forms of behavior).	enable Cl. to feel worthwhile and that Cl.'s expressions about himself are important. C. serves as a model for Cl. rather than as a manipulator of Cl.'s behavior.
LEVEL 5 C. shows unrestricted thoughtfulness, kindness, and consideration for Cl.'s worth as a person, no matter what Cl.'s beliefs and feelings may be. Thus Cl. feels free to be himself, even if this means regression, defensiveness, or disliking and rejecting C. C. attends to and shares Cl.'s happiness, hopes, and successes as well as his depressions, despair, and failures. C. makes one requirement of Cl.—work; that is, Cl. is expected to communicate personally relevant material.	C.'s verbal and nonverbal expressions indicate appreciation, genuine pleasure, and deep esteem for Cl.'s feeling and experiences. C.'s recognition of Cl.'s potential to handle his own concerns encourages Cl.'s problem solving and decision making as a free agent of his own affairs. C. recognizes Cl. as *somebody* of importance to himself, to C., and to others.

*C. = counselor; Cl. = client.

SCORE SHEET FOR WARMTH SCALE

Place check under *one* of the levels for *each* characteristic. The warmth level equals the total number of checks under each category divided by 2, the total number of categories. This is the average level of all the characteristics.

EXAMPLE: Positive regard   Level 3
          Respect   Level 2
                  Total = 5
        Divided by 2 = 2.5 (level 2+)

Characteristics	Levels				
	1	2	3*	4	5
Positive regard					
Respect					
Total No. of checks under each category					

*Minimum facilitative level.

Warmth scale (Table 7)

Two factors are included in this scale—positive regard and respect. The items for the scale for degree of warmth are based on those of Truax and Carkhuff (1967) and Carkhuff (1969).

Each person shows regard for things and people; individuals take notice of things and people to some extent, evaluating them favorably or unfavorably. Insofar as people are concerned, the more positive or favorable regard one person feels for another, the more likely this person would be to take into account the other person's needs and give attention to the satisfaction of these needs. However, heeding the call for help is not quite the same as respect. Respect goes one step beyond regard and adds another dimension that distinguishes each individual as unique and recognizes each individual for his particular potentials.

There is a fine line between regard and respect since their characteristics do overlap. One more distinction, nevertheless, may be stated. Regard may be either positive or negative. One person may consider another (regard) in an unkindly, unconcerned, uninterested manner (negative regard) or in a kindly, concerned, interested manner (positive regard). Respect, on the other hand, can only be positive and varies only in degree and not in direction of negative or positive. Therefore both regard and respect are included in the characteristics of warmth to provide a fuller picture.

BEHAVIORAL CHARACTERISTICS

Which is more important, the content of what one says or the way it is said? Which is more important, the message or the massage (McLuhan, 1967)? There probably is no answer that would apply to all people, about all kinds of situations, in all settings. The most effective balance between verbal and nonverbal elements in an interpersonal encounter varies on a person-to-person basis. Different studies of the relative contributions of verbal and nonverbal behavior to the functioning level of empathy, genuineness, and warmth have come up with conflicting results as to which is more significant.

Verbal cues

Verbal cues refer to the words used and the sounds made. The intervention of words may help or hinder the smooth flow of the counseling transaction. The effective counselor rarely gets into the "word act." When he does, it is to encourage the client to do some more self-exploration about how he contributes to his problems. The counselor tends to push for "here-and-now" words that refer to the present situation rather than the past. Thus the counselor uses words to keep the client from straying too far from his immediate behavior or the present problems.

Tonal qualities have a wide range of possibilities that may indicate positive or negative feelings. The tone of voice may sound close or distant, loud or quiet, tense (strained or tight) or smooth. A tone that is close, quiet, and smooth is more likely to suggest a higher level of empathy, genuineness, and warmth than a tone that sounds distant, loud, and tense. Emotional responses are also conveyed by such sounds as sighing, crying, coughing, and laughing.

Nonverbal cues

Verbal cues (the words used and sounds made) give only a partial picture of a message. Words in particular, because they are so dependent on individual interpretations, may distort rather than assist in understanding. Body language and other nonverbal behavior are learned during infancy long before words become part of the child's behavior. As the child develops, cultural demands step in and shape the gestures and body movements according to the prevailing customs. More personal influences of the people caring for the developing child also enter into the expression of nonverbal behavior. Since nonverbal behavior is more revealing than words, perhaps individuals condition themselves not to show their feelings through facial expressions. The "poker face" is a good example of concealing emotions.

In spite of the attempts to control the overt display of nonverbal behavior, feelings are not so easily squelched. There are numerous channels through which feelings manage to be revealed to the observant counselor. Similarly, the client may "feel" the attitude of the counselor by observing the counselor's nonverbal language.

Gestural movements. Individuals gesture with their arms, hands, legs, and feet. They may move their arms and legs up and down or from side to side. The finger-tapping, knuckle-cracking, leg-shaking individual shouts forth his tension, impatience, and displeasure. The open, accepting individual is more likely to have his arms open, his legs uncrossed, his fingers still, and his hands loosely clasped on his lap. Clients express more liking for the counselor whose knees are steady and who is from 3 to 6 feet away.

Head movements. The constantly moving head or the stiffly held, erect head has negative values for the client. Greater warmth, genuineness, and empathy are exhibited when the counselor periodically nods his head up and down or holds his head tilted to one side in a "listening" position. Clients receive more acceptance from the counselor when he holds his head slightly forward rather than backward or bent down.

Facial expression. Facial expressions are probably the most frequently observed cues. The smile of the warm and empathic counselor is much more effective than the tight-lipped, frowning counselor. Eyebrows have varied possibilities for expressing disbelief (one eyebrow raised), surprise (both eyebrows raised), or concern and/or distress (both eyebrows raised with creases in the forehead). The meaning of the movement of the eyebrows must, of course, be interpreted within the context of the client-counselor relationship.

Eye contact. The significance of eye contact in interpersonal relationships is demonstrated by the reactions of disturbed individuals. The more uncomfortable an individual feels with another person, the more likely he will avoid looking at him. Paintings of people done by severely emotionally disordered individuals often show the eyes in a bizarre (odd or absurd) manner.

Eye contact differs, also, depending on the status of the individuals involved. Individuals who are perceived as being of a higher status are afforded more eye contact than individuals considered to have a lower status. The student is likely to look at the college professor more frequently than he would look at a graduate student assisting the professor. Dependent individuals, seeking self-status, are inclined to reveal a relatively high level of eye contact as part of their approval-seeking behavior.

For the effective counselor the direct, occasionally changing gaze is evidence of a higher functioning level of the characteristics of the Truax triad than staring, glaring, squinting, or turning his eyes away from the client.

Eye contact serves another function. Changes in eye contact and/or body position appear when verbal exchanges are beginning or ending. In fact, counselors may employ these nonverbal cues as ways in which they alter the course of the discussion.

Body orientation. Constant movement or constant rigidity is a less desirable non-verbal characteristic. The comforting, understanding counselor tends to turn his shoulders and legs toward the client and lean forward slightly in a relaxed position. Con-

trasted with the rocking motion and the sideways position, the counselor's face-to-face position demonstrates a greater closeness with the client and also a less tense, more concentrating attitude.

Physical space. The preferred distance between individuals has cultural implications. There seems to be a graduated preference for distance that varies with the culture, with the social situation, and with men and women. People in the United States are generally inclined toward closer distances than in some European cultures, but American women are much more cautious about distances in physical space than are men. For the counselor the importance of physical distance depends on the client's functioning level. Being too close too soon may result in the client's confusion and embarrassment.

In general a closer distance between client and counselor indicates a more positive counselor attitude. However, the empathic counselor is aware of the appropriate distance for client comfort. Exercises 8-8 and 8-9 offer experiences in sharpening awareness to bodily cues and to the use of physical space

Exercise 8-8
TUNING IN TO BODILY CUES

The goal of this exercise is increased awareness of body cues and the feelings associated with these cues. Two circles are formed. Five trainees seat themselves in an inner circle so that they face the trainees in the outer circle. Thus, depending on the size of the outer circle, there should be two or more trainees observing each trainee in the inner circle.

DIRECTIONS FOR THE INNER-CIRCLE TRAIN-EES: Seat yourself comfortably so that you face those in the outer circle observing you. Close your eyes. Sit quietly for a minute. Then describe to the outer-circle trainees what you feel and how you feel it. After the first trainee completes his description, the second trainee to his left describes his feelings. This continues until all five trainees have completed their descriptions.

The next time each inner-circle trainee is a counselor. He selects someone in the outer circle observing him and walks over to him. The inner-circle trainee extends his hand to touch this person, his client; he describes to the client what he feels and where he feels it. The trainee to the left describes his sensations as soon as the trainee who is speaking finishes and sits down. The procedure continues until all five inner-circle trainees have completed it.

DIRECTIONS FOR THE ENTIRE GROUP: The five inner-circle trainees join the larger group and a general discussion is held of the recorded observations. Comparisons are made of the differences between verbal and nonverbal behavior for each trainee when his eyes were closed and when he role-played a counselor with his client.

Exercise 8-9
PHYSICAL SPACE AND
DISTANCE BETWEEN PEOPLE

Draw stick figures of yourself and the following persons:

1. Your father
2. Your mother
3. Your sister and/or brother
4. Your teacher and/or employer
5. Your friend

Draw the stick figures of yourself and the other persons listed as you usually feel when talking to each person according to size (smaller than, the same size, or larger than the other person), physical distance (near or far from the other person), and position (body facing toward or away from the other person).

After you complete the five stick drawings *as you usually are*, draw five more representing yourself and the same people, but this time make their sizes, and the positions and distance as you would like them to be.

What have you found out about your relationships with the five people of whom you drew pictures as you usually are? How do you feel the space around you and the distance between you reveals your interrelationships? What have you discovered about how you would like your relationship to change?

INTIMACY

It is time to put things together, to find out as Alice asked at the beginning of this chapter ". . . which way I ought to go from here?" This road to "somewhere" has three large billboards.

Development of identity

Each person begins as part of someone else and must move away before he becomes involved with these origins again. In other words, each individual must detach himself from his parents and become himself before he can view his parents as people and become involved with them at another level.

From attachment to detachment

The unborn child is an obvious example of physical attachment within his mother's womb. This attachment continues after the child's birth but with an additional twist— the newborn infant must learn to become dependent on others so that he may obtain satisfaction of his needs from them. The infant must, in fact, train the people around him to answer his cries and gurgles. A form of emotional attachment begins that is closely interwoven with physical attachment and satisfactions. Out of this arises trust or mistrust, which is the beginning step in the development of identity. For the infant, trust signifies that this is a fairly predictable and mostly pleasant world. Perhaps, the infant also feels, "I am important to someone." This feeling of importance, of belonging, of being cared for, of dependence moves the infant into self-exploration and into mastering himself and the things and people around him. The infant begins to separate himself and to become differentiated from the blurring confusion around him.

It is interesting to note the parallel of the infant's progress with that of the individual who seeks or is brought for help. The individual finds the answer to his problem, whether it is large or small, with the assistance of a more knowing person. Therefore the seeking person must feel that he can trust the helping person before any changes will occur. The seeking person attaches himself to a counseling transaction in which he feels that he is important to someone, belongs, and is cared about. With this dependence on a helping person, the client is better prepared to explore and to master himself and the things and the people around him. The confusion becomes explainable and new relationships are initiated in which the client detaches himself from the counseling situation toward independence. Every so often during this developing period the client returns for emotional refueling and starts out again with renewed vigor. Just as the crawling infant acquires steadiness to stand and to walk, so is the client able to begin to function on his own.

At first the infant or toddler requires the physical and psychological nearness of his mother to develop self-esteem and autonomy (independence). Later, when he begins to move about on his own, he may become overenthusiastic and temporarily dismiss the frustration that comes from failure to get to where he is heading. It is at this time, when the toddler is venturing farther away, that he needs his mother's warmth, understanding, and support the most.

The comparison of the infant's striking out on his identity adventure with the client's activities in establishing or reestablishing his identity is apparent. The person seeking help is more or less trying to define or redefine himself. The college student puzzling with the counselor over what should be his college major and the alienated, hostile, or depressed individual are both trying to refine their identities. The college student wants to discover which major fits his identity best and hopes to discard some of the choices (detach himself from them). The pressure of his parents or of the counselor to fit the college student into their molds encourages attachment or aggressive detachment and a blurred identity for the student. The student needs warmth, understanding, support, and also the opportunity for self-direction in defining himself and his goals more clearly. This process is rapid for some students and proceeds more slowly for others.

The alienated individual is enveloped in a more serious identity crisis. Such an identity crisis may occur during any of various change points in an individual's life. At each of these points the person is making another shift from attachments to certain characteristics and/or people to detachment. The first of these points occurs when the infant physically leaves his mother as he begins to walk. The toddler detaches himself more and more as he develops confidence in his increasing ability to stand on his own, both physically and psychologically. Away he runs, only to return periodically to make certain of his mother's presence if he should need her comfort. If the mother should leave the room so that the toddler cannot find her, his activity slows down, and he shows signs of alienation from his surroundings as he decreases his searching, his moving about, and his interest in his surroundings. It is almost as if the infant were saying, "It's all right for me to leave you [to become detached], but I need you around so that I can be me."

Several other change points result in identity definition for the individual. Adolescence, for instance, brings with it the even greater change point. The adolescent must adjust to a physically changing body as well as to the push/pull of detachment from his parents and attachment to his peers. This is a period when identity conflicts with anonymity. The adolescent pushes away from the viewpoints of his parents in an effort to establish his separate self-identity; yet he still needs some important others to whom he may relate. So he turns toward the peers he selects and identifies with their codes of behavior. He adopts the life-style, the hairstyle, and the uniform of the day, becoming more anonymous in his peer crowd. He eventually resolves this clash between identity and anonymity if the social situations allow it.

This clash becomes more unsurmountable for the individual from minority groups. The black youth finds that in spite of the new enlightenment, to many he is still "Boy," openly stated or secretly practiced. To ensure his identity the black man must choose from several options. He may detach himself from a black identity and become more like white individuals or he may become more militantly attached to his black identity with subsequent detachment from "whitey." A third way is a little more subtle—he becomes hostilely seductive, particularly with white women.

The Chicano must phase out his Spanish-speaking identity and in the process assume a slow-learner identity in the English-speaking school. He, too, has several choices for his resolution of the clash between identity and anonymity. He may detach himself from his parent's customs and speech, his Chicano friends, and their life-style and make a show of his Anglo identity. On the other hand, the Chicano, like the black or the Indian, may adopt a militantly Chicano identity in which he detaches himself from the Anglo world, speaks Spanish, leaves school, and attaches himself to the hairstyle and clothes-style of his Chicano peers. Out of these unresolved conflicts rises identity confusion or a constant battle for identity survival.

What does all of this mean to the counselor? First of all, it means that the counselor must experience accurate empathy. This understanding must begin with himself. Does the counselor fit the client's needs? Does the counselor know himself well enough that he is able to be genuine in his responses to the client, no matter what his race, color, or creed? Does the counselor trust his own identity so that he may encourage the client to trust him? Has the counselor progressed from the attachment/detachment conflict to the next stage of progress, detachment/involvement?

From detachment to involvement

Christopher Robin (Milne, 1924) brings to attention his growing independence when he says:

> I never did, I never did, I never *did* like "Now take care, dear!"
> I never did, I never did, I never *did* want "Hold-my-hand";
> I never did, I never did, I never *did* think much of "Not up there, dear!"
> It's no good saying it. They don't understand.

Independence is not easy to achieve for the growing child, for the struggling client, or for the beginning counselor. Each of these individuals forms a strong attachment to an important other person. Each of them becomes dependent (attached) to this person, whom they learn to trust and to look toward for help. Yet each of these individuals must also become independent (detached). The child wants his parent to let him walk alone, near the parent, perhaps, but without holding hands. The client and the counselor also need to be gradually released from the symbolic hands that guide them.

As the child, the client, and the counselor-trainee begin to form their own ideas of self-identity, they may be pleased with the care, the direction, and the outstretched hand. However, if the development stops at this point, complications arise from the continuing attachment. Instead of self-identity, smudged carbon copies result. The individuals are not as confident or as competent as the original.

In order to move from detachment/independence to involvement/interdependence the individual must feel comfortable with himself and with others. By detaching himself the individual is able to become more aware of himself, more alert to his ability to achieve, and more defined as an individual. He finds his meaning for life; he becomes involved.

The alienated and/or depressed person reacts differently. He cannot move from either severe detachment (alienation) or deep attachment to self-concern (depression). Unable to find joy or meaning, he may strike out for power, excessive pleasure, or even death. None of these goals leads to self-satisfaction, for the individual's goal is really frustration. The alienated or depressed individual suffers from what Frankl (1963) called "hyperintention." Such excessive attention to his own and/or to others' needs is not the same as involvement. Only the person who accepts the challenge of first fulfilling himself can become involved with others. The alienated or depressed individual is often tormented with attachment or involvement as a threat to the security of his withdrawal.

As an individual becomes more aware and acceptant of his own weaknesses, strengths, and imperfections, then he can move from the loneliness that detachment imposes to the satisfactions that involvement and intimacy bring. Exercise 8-10 focuses on searching for some of these strengths and weaknesses that may be stumbling blocks to openness.

Exercise 8-10
STRENGTHS AND WEAKNESSES

Complete the chart on p. 192 and then gather into a small group of no more than five people to discuss your findings. You, of course, have the right to refuse to discuss anything you write; so feel free to write comments for yourself even if the information is not to be shared with others.

In the first column is a list of people. Write next to each person what you like most (his strength) and what you like least (his weakness) about the person. In No. 8 and No. 9, fill in two other people and complete the columns for them.

After completing the list, look it over to see if you notice any pattern emerging. Are your strengths (what you like most) and your weaknesses (what you like least) similar to anyone else's?

Alienation is an example of another kind of involvement—self-involvement. The alien-

	I LIKE MOST ABOUT	I LIKE LEAST ABOUT
1. My mother		
2. My father		
3. My sister		
4. My brother		
5. My friend		
6. My teacher		
7. My employer		
8. My _____		
9. My _____		
10. Myself		

ated individual is detached from others while he hides in himself. This self-attachment is a protective device to conceal the anxiety he feels in person-to-person encounters. Such an alienated person stops growing outward, toward others, and grows inward toward himself. His self-concentrated behavior is revealed in the way he handles his relationships with people. The self-concentrated individual evaluates other individuals as "better than," "not better than," "more important than," or "less important than" he is. He finds ways to exploit situations and people in order to satisfy his own needs. Rather than being interested in a partnership with another respected individual, he persistently seeks gratification of his own desires; yet he remains unsatisfied. He dehumanizes the other person and thinks of and feels toward the other person as an "it," an object. Dehumanization and anonymity are close allies. Anonymity results in treating another person as a formless "whatchama-call-it." "What's his name" takes up physical space but very little psychological space. "Who's that" has no identity as a human being; he is anonymous. Why pay attention to a nonessential except to move him out of your space?

Dehumanizing another person makes violence and even murder possible and almost inevitable. Violence does not need to be physical to be disturbing. It may be the stealing of psychiatric records for political reasons. It may be labeling someone from a minority group a slow learner or mentally retarded. Both of these do violence to the identity and personal growth of an individual. Murder takes many forms. It may be the killing of the anonymous enemy during war, the sniping of individuals from the rooftop of a building, or the slow torture of hunger and malnourishment.

". . . The only essential requirement for counseling is humanness, which is the capacity of an individual to be open to another human being without judging and without losing his own individuality" (Webster, 1973). Humanness includes an individual's weaknesses as well as his strengths. Involvement of the counselor's humanness in the counseling situation therefore requires that the counselor feel comfortable with sharing his human qualities. The primary objective of this sharing is to help the client accept himself "as is" so he may start becoming what he wants to be. There is a catch in this sharing. The counselor must be sufficiently detached from his own desires during the counseling transaction so that he does not become self-involved. The counseling help is for the client, not for the counselor.

The attached (dependent) infant grows into the interdependent adult because of the interest and caring of others. He moves ahead to self-respect because of the intimacy others feel and experience with him. He learns responsibility because other responsible people are involved with him. Caring means both love/liking and restraints/discipline. Stunted psychological growth often results if the developing child, client, or counselor is exposed to only one aspect of caring, exclusive love, or exacting discipline. These characteristics of interdependence, caring, discipline, involvement, and responsibility are essential for the effective counselor.

What an individual likes most and likes least about himself give some cues to self-feelings. Exercise 8-10 sought these like/dislike responses. Exercise 8-11 takes a look at self from the angle of "needs," "wants," and "changes."

Exercise 8-11
NEED, WANTS, AND CHANGES

Complete the following statements and then gather in groups of five to discuss your feelings. You may refuse to reveal anything you have written. The statements you make must be in terms of "here and now," the present, not the past or the future.

WHAT I NEED MOST: (Interpret the difference between need and want according to your own meanings.)

WHAT I WANT MOST:

I WILL CHANGE BY DOING THE FOLLOWING:

I DON'T THINK I CAN CHANGE BECAUSE:

Aspects of intimacy

Intimacy is more than closeness. Intimacy also demands the acceptance of the worthiness of others and commitment to a relationship with another human being. Close association, self-sharing, inclusion, and a kind of cohesion (sticking together) are all part of intimacy.

Unfortunately the word "intimacy" is often misused to mean that someone has some contact with someone else. This contact may be called "intimate" because of the secret thoughts one person reveals about himself or about another. Often these "secret" thoughts are selected carefully to make points, not for the sake of honesty. Intimacy is more than knowing *about* someone conversationally or sexually. To know another person intimately there must be an exchange of thoughts, a close and confidential relationship of trust.

Exercise 8-12 explores some of the feelings associated with intimacy.

Exercise 8-12
ATTACHMENT, DETACHMENT, AND INVOLVEMENT

Select the word from the following list that best describes your feelings for the persons listed in the chart on p. 194.

Attachment	Caring
Detachment	Responsible
Involvement	Isolation
Relatedness	Alienation
	Intimacy

If your feelings are positive (pleasant desirable), write the selected word under "most of the time" or "some of the time" in the "positive" column. If you think your feelings are negative, place the word under "most of the time" or "some of the time" in the "negative" column. For example, if you feel negative (unpleasant, undesirable) attachment toward your mother most of the time, write the word you selected under "negative, most of the time" on the line following the word "mother." In No. 8 and No. 9, list two other people and select the words that best describe your feelings for them.

What does the listing and words tell you about yourself and your relationship with other people? Compare your responses on this list with your responses in Exercises 8-10 and 8-11.

I FEEL FOR MY:	POSITIVE		NEGATIVE	
	Most of the time	Some of the time	Most of the time	Some of the time
1. Mother				
2. Father				
3. Sister				
4. Brother				
5. Friend				
6. Teacher				
7. Employer				
8. ————				
9. ————				
10. Myself				

To know another person intimately there must be not only an exchange of thoughts but also a close and confidential relationship. Intimacy takes many forms. The more of these forms an individual honestly expresses to another individual, the more complete is the intimacy of the relationship.

Cerebral intimacy

The easiest intimacy to accomplish is probably the cerebral kind. It only takes the right choice of words. Keeping oneself encased in one's skull prevents others from getting too close and in this way maintains a safe distance for the skull-hugger. The magic of words lulls many individuals into a false feeling of togetherness. "Aw, come on, you know you're rationalizing; you know you really dig this stuff" has the sound of intimacy but is actually manipulation.

Cerebral intimacy may also be deceitful. Admitting one's deceit to others and, most importantly, to oneself is a difficult task. It is much easier to cover up with candy-coated lies and to create deceptions to conceal real intentions. To conceal real intentions from oneself and others may be a plan of action or manipulation.

Sincerity means one has to face oneself first. The false compliments with which one person flatters another, the mechanical words of praise, may wear the mask of intimacy but is just extravagant cerebral game-playing. A dialogue can be conducted only between people of equal worth. This does not require that the individuals be equally intelligent or equally educated, nor does dialogue prevent angry expressions. Sometime the dialogue turns to duelogue (p. 130). As long as the cerebral battle is fair, frank, and respectful, the individuals involved are on their way to intimacy. The two counselors in Example 8-2, Mr. V. and Mr. P., are on different trips and the client, Peter, bounds forward or springs back in accordance with their different approaches.

Example 8-2
TWO COUNSELORS—ONE CLIENT

1

MR. V.: Hello, Peter, nice to see you. How about the red chair? It's very comfortable. Say, aren't you the brother of. . . .

PETER: Hello, I—well, it's been a hard year for me. I guess you know my brother, Jim. We . . . he managed okay, even got a scholarship to college. Everything was okay with me until this year. Ben and I, we. . . . Is this confidential?

• • •

MR. P.: (*Notices Peter's peculiar stride as he walks in on the balls of his feet with his hands jammed into his torn, black leather jacket. Hair hangs around Peter's forehead and face. Mr. P. stands and walks to Peter.)* Hi.

PETER: Huh, well—phew!! This is a real drag. My brother, Jim, told me about you. Remember him? Won a scholarship. Until this year, everything was okay. Ben and me, we. . . . *(Silent.)*

2

MR. V.: Oh, yes, Jim—uh, mean Peter, of course, I remember your brother. Brilliant student, used to come in to chat with me, played guitar, too. Sure, remember him well. Hmmmm, so what's the problem?

PETER: *(Silent.)*

• • •

MR. P.: *(Moves his chair slightly forward and leans toward Peter. Looks at the deep lines between Peter's eyebrows and Peter's clenched hand rigidly pressing his thighs.)* Kinda hard to begin?

PETER: Mmmmm, very. So tight; how do you . . . ?

3

MR. V.: Come now, Pete, you can talk freely here. Trust, the very essence of inter-personal relationships, trust is an ever-lasting credo [belief] you must establish. The relief that comes from catharsis—you know, tell about it; get emotional release. Helpful, very helpful.

PETER: Mr. V., I dunno. Wonder if I can do it alone. Ben will get uptight if he finds out I've been talking to you.

• • •

MR. P.: Peter, seems like heavy stuff you're carting. How about starting with when you met Ben.

PETER: Ben, yes. Met him last May, no, April, at a pot party. Saw him after almost daily. Funny, never noticed him before, but suddenly he seemed to be around the same places, and then. . . .

The trend of the conversation in Example 8-2 is apparent. Mr. V. is the skull-hugging individual whose pseudointimacy serves his own ends. Mr. V. delivers words but little empathic understanding. Peter finds re-sponding to Mr. V. too difficult. Mr. P., on the other hand, indicates by means of body language (moving and leaning toward Peter) and his few well-chosen words that he is "in there" with Peter.

Words do indeed help to establish inti-macy. However, overindulging in words, supercerebral intimacy, can be a front for alienation and/or hostility rather than hon-est intimacy. Social intimacy is another channel to increase or decrease closeness.

Social intimacy

Social intimacy is similar to cerebral inti-macy insofar as it varies in intensity and the way it is expressed. An individual may be so eager for other people that he constantly seeks the company of others. Interpersonal relations become the means of his survival. The question arises as to whether this is intimacy and openness or a sign of uncon-querable loneliness.

On one end of the scale of social intimacy is the person who hoards his privacy and keeps to himself. On the other end is the individual who is almost unable to exist without the company of others. In between is the person who feels free enough to either open his doors to others, to be prepared to share with others, or to ask for and arrange for privacy without feeling apologetic. There is a difference between what Berne (1964) calls pastimes or games of intimacy and game-free intimacy. Pastimes are involved with a special kind of cerebral intimacy in which special words are used for certain events to achieve surface social intimacy.

DON: Hey, man, did you hear about Butch? He finally made it.

HAL: Butch? Oh, yeah, that dude; so he marked one. Whaddyaknow?

The ritualistic language, the noninvolved discussion, point to intimacy that hardly scratches the surface of interpersonal trans-actions. One more step into social organiza-tion and games becomes the way of things—and people. Sometimes the games become grim in the attempt to be intimate. The individuals advance into a little more in-volvement and plan a little more carefully. However, the behavior is commanded by rituals so well regulated that spontaneity is lost. Therefore cerebral intimacy must be controlled and social intimacy carefully pro-grammed. The rigidity of contacts makes the social intimacy unreal. Friends are selected according to their ability to play along in mutual gamesmanship.

MR. C.: Yes, indeed. I enjoy these swinger groups. Jealousy? Of course not at all. My wife still is faithful to me. I've really learned a great deal about myself, even helped in my sexual relationship with my wife.

MS. L.: Know what you mean. Freedom. Manny and I, we know ourselves so well and we've gotten closer, much closer since we joined the group. How about moving along now?

Social intimacy implies mutual availability and mutual acceptance. This does not signify that each individual is everlastingly available to the other person or even everlastingly acceptant of the other person. It does mean that privacy is expected and practiced. It means that there is genuine cerebral intimacy. Ritual social ability and ritual avoidance do not fit into genuine social intimacy.

HELGA: Bob, it won't work; just can't get into this housewife bit. Feel guilty about your taking over.

BOB: Noticed it today, how tense you look, so uptight. I really feel where you're at. We've just got to get into this thing together.

HELGA: Well, you do . . . you are helpful, yet—wonder if my guilt is blown up because I keep sensing your annoyance. Are you really telling me like it is? Keep wondering.

BOB: Hell, Helga! You know damn well I don't like getting into the house scene any more than you do. But maybe I dislike it just a little less while you're getting into this new thing of yours. Let's not fight it; we'll readjust later when there's less static for you, okay? (*Walks over to Helga and puts his hands on her shoulder and then embraces her gently.*)

The conversation between Helga and Bob brings forth some feelings of anger as well as tenderness. Yet the prevailing mood is intimacy, which is honest in Bob's cerebral, social, physical, and emotional responses. Bob's movement toward Helga went one step farther than cerebral and social intimacy to physical intimacy.

Physical intimacy

Intimacy is considered most often from the perspective of physical closeness. Perhaps it is more accurate to view the depth of intimacy on this physical level. Morris (1971) argues that physical intimate behavior is crucial to human existence. This argument is supported by others who have studied infants deprived of physical contact and found that they waste away without this kind of contact comfort. This wasting away occurred even though the infants were provided with adequate nourishment, rest, temperature, and so on. When certain people were assigned to rock and otherwise hold these infants, they began to thrive (Ribble, 1955; Spitz, 1965). Although these studies concentrated on infants in institutions, hospital personnel and pediatricians have discovered that physical contact is an important survival factor from the moment of birth. As with all forms of intimacy, physical intimacy begins with the parent being physically intimate with the infant and goes on to the child's acceptance of self-intimacy and gradually physical intimacy with others. The individual who feels comfortable with his own body is an "embodied self" (Laing, 1960). He feels his physical self together as an organized whole. Individuals who act as if they are disconnected from their body (disembodied) reveal this feeling in the way they talk about themselves, in the way they carry their body, and in the way in which they use space. The person who rejects all or part of his body or fears and/or is ashamed of the consequences of body contact speaks with displeasure of his physical dimensions, his shape, or his odor. He may tackle his distaste by calling verbal attention to his body with hostile humor or even sometimes with pretended affection.

Furthermore, the individual who feels that part or all of his body image is distasteful or even disgusting may reveal these disquieting feelings by the way in which he seems to detach himself from his body. He may show this detachment in several ways:
- By the lack of care for his clothing and/or his physical cleanliness
- By the thrusting walk that gives the impression of the head, shoulders, trunk, and legs moving at different speeds as if these body parts were not joined together
- By rigid, robotlike movements

Another response that demonstrates the individual's negative body feelings is the way he uses space. Hall (1959) discusses man's use of space in his book *The Silent Language*. He points out that the ranges of distance may be divided from the "very close," which is the "soft whisper" of from 3 to 6 inches, to "stretching the limits" or "hailing distance" of from 20 feet indoors to 100 feet outdoors. Interaction distances vary in different cultures and from individual to individual. An individual who rejects his own body is more than likely to reject the body of another person. He keeps his distance by manipulating the physical space between himself and another person. He places himself sufficiently away from the other person to require a louder voice and to discourage touching. He may pad his body with enough poundage that his body contours maintain distance from others. Limiting contact by distance hampers physical intimacy.

Physical intimacy varies from person to person. The degree of closeness an individual can comfortably handle is revealed in the space an individual arranges for conversational purposes. Withdrawn individuals are more likely to select distances farther away from other people. A rapid departure frequently results if someone invades the spatial safety limits arranged by the withdrawing individual.

Emotional intimacy

Cerebral, social, and physical intimacy are less threatening than emotional intimacy. Some people are terrified by emotional intimacy and fear that exposure of their feelings may make them so vulnerable to themselves and to others that they will be unable to deal with the results of their emotional openness.

It is not easy to establish emotional intimacy. Many individuals build up an elaborate system of game-playing to avoid emotional intimacy. Aggression and even more serious consequences often result from the inability to be close and open with others. It is simpler for those who avoid closeness to sprinkle their lives with "ought's" and "should's" rather than with honest commitments to themselves and to others. Overconformity smoothly destroys individuality as well as emotional intimacy. How can one give a genuine response to another person if one becomes dependent on the other person for perpetual approval? Sometimes the individual decides that the only way to break away from such dependency is through cruelty and insensitivity to the other person's feelings.

Emotional intimacy suggests two persons who are tuned in to feelings with firm roots of identity and healthy growths of togetherness. It suggests saying "we" or "us" rather than "they," "them," or "that." It suggests pain, pleasure, and also sharing. It suggests waves of warmth that are nonpossessive of the other person. It suggests a continuous growth of these characteristics, not a level reached—striving, not finishing.

Fulfilling intimacy means all of these characteristics—cerebral, social, physical, and emotional. There is no absolute level of intimacy but rather a matter of degree. The burden of phoniness, however, is too much for intimacy to carry. Phony intimacy is paid for with lies and concealment.

Each individual asserts his own personal style of emotional intimacy. This style of intimate behavior (behavioral effects) develops from an individual's background influences, his psychological responses, and situational features of his immediate environment (p. 59). Out of these influences, identity is shaped and intimacy comes. Commitment and engagement with others is a test of one's identity. If an individual is shaky about who he is and about his meaning for living, he is also apt to be strained and reserved in his emotional transactions with others. He cannot abandon himself to the joys of emotional sharing if he is bewildered by his vague identity.

Intimacy's opposite is separateness, remoteness, isolation, and alienation. These signs of distance encourage destruction and/or ignoring of others who may be considered dangerous to one's identity-seeking. Otherwise a desperate pseudointimacy leads this same person to try to merge his shaky identity with another adult, a guru, a cause, or drugs. If none of these half measures work, a vicious cycle of searching and retreating of excessive trusting or mistrusting begins.

The counselor must look to himself and function on a higher level of authentic intimacy. He is free to feel empathic, genuine, and nonpossessively warm. He is free to begin self-evaluation by means of self-disclosure.

Self-evaluation through self-disclosure
Looking back: a summary

Many thoughts ago these pages began with the suggestion to "Look and listen for the way to go." A summary of the steps taken to get to this point should advance this journey to "somewhere."

This book is about the undergraduate training of beginning professionals in nonmedical human services. The first step toward becoming this beginning professional is self-examination. Self-understanding leads to awareness of prejudices and other hangups that might interfere with the helping role. These human service workers are agents for change who give to others the same opportunities and rights they have and strive to get for themselves. Their crucial function is to help others to help themselves.

During this process of self-understanding and attitude exploration there should also be a gradual building up of knowledge and skills. The acquisition of knowledge and skills starts with learning to observe verbal and nonverbal behavior as well as to record and to report these observations (Chapter 2). Observation is the foundation on which all other skills are built.

The "what" and "how" of interviewing (Part 2) stress getting out of the client's way so that the client may draw on his own resources and make his own decisions for change. Three types of interviewing relationship—action/reaction, interaction, and transaction—are described. The transaction between the interviewer and the client is significant for establishing accurate understanding (empathy), honesty and realness (genuineness), and unrestricting closeness (nonpossessive warmth).

Four determinants that influence behavior (Chapter 4)—background roots, psychological bases, situational features, and behavioral effects—supply knowledge about human behavior that is important in opening, developing, and ending the interview (Chapter 6). These three phases of structure exist for all interviews whether the interview goals are information oriented, behavior modification, or experiential (Chapter 5). Skill in managing the beginning phase, in encouraging the client toward self-exploration, and in moving with ease into termination of the interview is how the interviewer helps the client find his way toward more self-satisfying functioning.

Interviewing is the core of counseling (Part 4). Similarities and dissimilarities among conversation, interviewing, counseling, and psychotherapy depend more on the professional attitude developed and the extensiveness and depth of knowledge, skills, and experience than on different kinds of transactions. The use of such expressive skills of communication as paraphrasing, summarization, reflection, confrontation, and interpretation (Chapter 7) requires additional practice under supervision before the helper may apply these skills effectively.

The three characteristics that are vital to the helping process are accurate empathy, genuineness, and nonpossessive warmth—the Truax triad described in this chapter. Scales are included for examining the functioning level of these three interpersonal skills. The result of a higher level of these skills is an intimacy that requires the openness of shared self-disclosures. The counselor discloses himself for the client's benefit; the client discloses himself in order to unravel his concerns.

Now: halfway down or up

The challenge now remains—halfway up or halfway down? Milne (1924) cleverly describes this dilemma in his poem "Halfway Down."

> Halfway down the stairs
> Is a stair
> Where I sit.
> There isn't any
> Other stair
> Quite like
> It.
> I'm not at the bottom,
> I'm not at the top;
> So this is the stair
> Where
> I always
> Stop.

Halfway up focuses on the present and the future; halfway down focuses on the past. "Up" or "down" is more than game-playing; it presents two different viewpoints. The trainee who is halfway down is looking at what he has learned and what is happening around him, and he is satisfied or at any rate slow to do anything about himself or the world around him. The trainee who is halfway up is aware of what he has learned but continues to search for more knowledge and increased competence. One person thinks he *has* become, the other *continues* becoming.

The following discussions contain some thought-starting notions for individuals who are halfway up.

Anti-shrinkthink. The simplest method of dealing with the protests of minority groups, women's lib groups, and radical/revolutionary/disenchanted people is to deny they exist. If one of these dissenting noisemakers happens to push his way into one's consciousness, all one has to do is to find that halfway-down stair, stop, and sit. Then it is easy to shut one's ears to unpleasant remarks about the needs of people or the impact of an unsatisfying and/or hostile society. Conversely, if one finds stopping on the stair uncomfortable and feels compelled to move up and out, he must then search for the sources of discontent and do something about them.

"Anti-shrinkthink" (Dilley, 1972) is one of the ideas proposed to make the helping profession more answerable to people and to their problems. It is a plea:

- For recognition of the need to support beginning professionals in their roles as helpers
- For a shake-up of the stereotypes about women's roles
- For a recognition of the prejudices against and pressures on minority group members and the poor
- For opposition to the label "mental illness" and to the indiscriminate use of any labeling
- For the right of every individual to find his own way toward satisfaction and self-fulfillment

Progressive counseling. How about counseling? Would the anti-shrinkthinkers find that most counselors are restricting conformists who repress their clients? Several recent critics of the counseling profession have presented it as reactionary (Banks and Martens, 1973) in spite of the fact that the counseling movement has typically supported change in the schools as well as in society (Adams, 1973). At present it appears as if the counseling profession is experiencing an identity crisis that is not easy to resolve.

On the one hand, counselors support clients in their right to select alternative life-styles and yet do not recognize or are uncomfortable with the need for significant changes in educational, social, and economic institutions to support these varieties of iudividual life-styles. Counselors are more prone to emphasize the need for change in their clients so that the clients may live more satisfactorily in society. The "Catch 22" in this effort is that if the client does learn to live in society, he loses some of his uniqueness. Yet, if these clients do not learn to live in society, they are likely to become outcasts, living on the fringes or gathering in isolated places to live differently.

Pragmatism, or practicality, is the moving force for many counselors. What is practical is too often measured in quantitative terms —how many students in a classroom or how many clients can be rapidly molded into accepting the educational, social, and political pattern planned for them. The basic assumption is that the problem is rooted in the client's inability to adjust. This opinion, which upholds the notion that one cannot change society, one can only change the person ". . . ignored one-half of the relationship of the individual and the society. Working only to implement the individual has the effect of giving implicit approval to the society" (Adams, 1973). The following ideas take into account both individual and societal responsibilities of the helping person.

- He must be able to explain his changing role to both administrators and to others.
- He should provide counseling and other assistance to special groups such as abortion counseling groups, drug crash pads, women's liberation chapters, and minority groups.
- He should interact with youth in the neighborhood settings.
- He should establish rap groups in shopping centers and other places that are accessible to people of various ages.
- He should become further involved in outreach activities such as organizing community groups according to community interests.
- He should develop and support introductory and continuing training programs for himself and other beginning professionals.
- He should, in other words, become what Carkhuff (1972) calls a "militant humanist," an agent for self-change as well as for social change.

Tie all of these ideas up and they result in an all-inclusive question: Does the helping profession dare to design a new approach to helping that is identified with human fulfillment rather than the patching of psychological wounds? Does it dare not do so?

Looking forward: self-evaluation

Although all of the ideas just mentioned will not be acceptable to all of the people in the helping professions, there is one suggestion that cannot be pushed aside—self-change for the helping professional. No helping person can hide on the stair at which he stops; he must go "up" to continuing education and accountability.

Continuing education. The knowledge and skills associated with the helping relationship are ever increasing and ever changing. It is the obligation of every person in the helping relationship to update his knowledge and refine his skills. Some of this new information is available from books and periodicals but a great deal more is gained through interactions with colleagues in workshops, conferences, and meetings.

In Maryland the Manpower Development and Training Division of the Department of Health and Mental Hygiene supported a continuing education program that was planned and organized by Mental Health Associates of Maryland, graduates from 2-year community college programs. The Mental Health Associates (MHA) of Maryland planned and organized a series of eight workshops with the assistance of Mary Gardner, a consultant from Manpower Development and Training Division. The overall program goal agreed on was as follows:

A program of continuing education in cooperation with academic institutions, consisting of orientation of new mental health associates and continuing education for all MHAs working in the Department of Health and Mental Hygiene, which will represent a combination of centrally organized and institution based learning/working experiences and which will be academically accreditable at the baccalaureate level.

The workshops concentrated on five skill areas of particular interest to the MHA participants—interviewing and assessment skills, individual counseling and therapy, family counseling and therapy, group counseling and therapy, and behavior modification. Each one of the eight workshops was organized around a particular client population—mentally retarded persons, drug-ad-

dicted persons, urban communities and minorities, persons who have experienced short-term hospitalization, individuals and families in crisis, and aged persons. "The Personal Orientation Inventory (POI) [Shostrom, 1962; see Appendix B] has been used to measure change in self-actualization or self-directedness as a personality dimension" (Gardner, 1973). Additional assessment was accomplished by feedback from peers and supervisors, two consultants who functioned as participant evaluators, and an evaluation session at the end of the series.

Continuing education may also be accomplished by study at institutions of higher education. This form of learning is worthwhile if the programs are planned to take into account the work experience of the beginning professionals and if the courses are planned in terms of the beginning professionals' needs. Academic credentials have set up artificial barriers to the advancement of beginning professionals. Numerous devices exist or can be established to assess the level of knowledge and skill of individuals working in helping roles and college credit may then be offered for this competence. These individuals would be more likely to become part of an ongoing educational process that recognizes their competencies and aids them in going on from their present level.

Accountability. Continuing education is actually one form of accountability. An individual fills the gaps and refines his knowledge and skills through some form of further education. Accountability usually signifies anything but self-evaluation. Writing reports, being supervised, and answering questions are the more frequent forms of accountability. These forms are at the far end of the measuring stick. The person who must do the recounting and reporting should start with self-evaluation before any other accountability is required.

Many individuals believe they are constantly evaluating themselves, and yet when they closely examine what they really are doing, they find that their evaluation is not planned but is often scanty and infrequent. The counselor self-evaluation scale on p. 201 is adapted from the original scale devised by Mozee (1972). The items may be altered to fit the particular circumstances of an individual's work setting.

COUNSELOR SELF-EVALUATION SCALE*

Key: 1 = Low   3 = Average
     2 = Below average   4 = High
          5 = Very high

	Level of competence					Level of interest				
	1	2	3	4	5	1	2	3	4	5
Counseling students Low ability										
High ability										
Physically handicapped										
Older										
Ethnic minority										
Counseling poor people										
Career planning counseling										
Personal problem counseling										
Group counseling										
Writing recommendations										
Conducting research studies										
Conducting in-service education										
Teaching regular classes										
Beginning human service workers Training										
Supervising										
Orienting New students										
New staff members										
Establishing communication with: Community										
High schools and colleges										
Other departments										
Making referrals										
Doing follow-ups										
Using counseling techniques Transactional analysis										
Behavior modification										
Client-centered										
Experiential										
Insight-oriented										
Your own brand										

*Adapted from Mozee, E.: Personn. Guid. J. **51**, 4:285-287, 1972.

Rate your level of competence in each of the areas listed; then rate your degree of interest in using that specialized skill or performing that particular task. In the left-hand margin, use a number scale from 1 to 5 to indicate the importance of the functions to you—1 is least in importance and 5 is most in importance to you. After you have completed your ratings, put together your thoughts about the self-evaluation scale into a summary statement that represents your present place in the helping profession.

After completing the self-evaluation form, the individual should explore whether he is performing the tasks he prefers the most, whether he is performing at an above-average level, and whether his summary statement discloses any areas that need to be examined further.

Accountability can be very threatening and result in defensive behavior. The critical feedback that concentrates on spelling out incompetence may lead to temporary changes; however, in order for accountability to serve as a vehicle for positive growth in competence, feedback must contain systematic, friendly, constructive comments of favorable as well as unfavorable performance. Accountability is not only for evaluating what is wrong with performance it is also for pointing out what is right. In addition, there should be a clear statement of the criteria and methods of measurement on which the evaluation is based. A vague remark such as "According to our observations you are doing very poorly," says noth-

ing constructive and eventually leads to a deterioration of performance.

Evaluation (accountability) may be focused on the client's goal attainment. If the client and the counselor arrange a contract, then they are able to use this as the basis for determining the goals that have been and that still have to be achieved. An example of a form for contract fulfillment in which the goals and the progressive fulfillment of these goals may be examined is given below. Any number of goals may be listed in accordance with client-counselor planning. Goals may be revised or added to during later counseling sessions. The counselor uses a contract for his own information or shares the completion of the columns with the client. Mutual accountability is effective with some clients.

Accountability is most fruitful if it offers suggestions and yet maintains respect for the individual being evaluated. The first direction that evaluation should take is self-evaluation, which may be used as the basis for reporting or just for self-information. Forms, scales, contracts, and follow-ups serve as devices for accountability. In addition to self-evaluators, outside consultants may be brought into the agency to perform the assessment. Accountability that serves the worker, the employer, the teacher, the administrator, the client, and the counselor is the most meaningful.

SUMMARY

Chapter 8 ties together the ideas of the book. There are two recurring words

| Name of client: _____ Date: _____ | | | | |
| Name of counselor: _____ | | | | |
GOALS SET	PRESENT ATTITUDE TOWARD COUNSELING	PROGRESS MADE	UNFAVORABLE REACTIONS	ATTITUDE TOWARD GOING ON

throughout this book—"understanding" and "acceptance." Both are part of humanness. From self-understanding (insight) emerges the openness, the self-disclosure, and the self-evaluation that are the foundation for empathy, genuineness, and warmth. These three interpersonal skills are the central ingredients for change. The empathic counselor exhibits both insight and outsight since he works at understanding himself as well as others. He looks at the world through the client's eyes and listens with the client's ears *as if* he were the client. He must experience closeness (warmth) with the client as well as remain alert to the messages the client sends. Observation of verbal and nonverbal behavior is the key to the counselor's understanding. The client "talks" with words, with sounds, and with his body. The counselor uses similar language to convey his messages, which are enveloped in empathy, enriched with warmth, and are trustworthy because of their genuineness.

Genuineness involves the counselor's willingness to be known. This ability to reveal himself to the client aids the client to untangle some of his feelings. The counselor tunes in to the client's needs so that he may know how and what ideas and feelings to share with the client. This sharing becomes a form of self-confrontation for both client and counselor. They both feed information about themselves to each other and also hear what they are saying. Thus there is not only mutual feedback but also self-feedback.

Empathy, genuineness, and warmth may be measured by scales that describe the different levels of verbal expressions. They may also be measured by observation of the kinds of words and nonwords (sounds and nonsense syllables) used. In addition, the language of the body, its movement forward or backward, gestures, physical distance, and eye contact are cues for determining the level of empathy, genuineness, or warmth.

The counselor's expression of warmth increases his personal attraction for the client. This personal attraction is advantageous since it encourages greater cooperation from the client. However, too much warmth, too much liking, too much attraction, and too much cooperation are apt to result in an overdependent client. If this should occur, the counselor must examine his motives—is he striving to satisfy his own needs by creating an overdependent client?

Out of warmth emerges intimacy. This freedom to feel and be close to someone hinges on the development of identity. An individual establishes a sense of selfness, of continuity, and of wholeness. Establishment of these physical and psychological boundaries of self is the springboard from which an individual can relate to others.

Identity develops from attachment (dependent) to detachment (independent) to involvement (interdependent). The troubled individual who comes or is brought to the counselor and the beginning counselor go through stages similar to the growing infant and child. They move from one stage to the next, gradually defining or redefining their identity.

Involvement takes at least two different directions. An individual becomes involved with others, interdependent, and intimate. Another individual becomes self-involved and either dependent on someone and over-intimate or independent and aloof from people. The individual involved with others is comfortable in approval-giving; the individual who is self-involved is often approval-seeking.

Involvement and intimacy are locked together. They both share the same body of characteristics that include acceptance, sharing, and inclusion. Intimacy includes involvement but delves deeper into feelings. There are several forms of intimacy; each of them alone leaves something wanting in the human transaction. These forms are cerebral, social, physical, and emotional intimacy. All of these forms put together make for a more complete human transaction.

The individual whose identity is assured, who feels comfortable with himself most of the time, does not consider his planning, changing, and learning as a finished product. He tackles not only the client's problems but also the societal problems that make it difficult for the client to express himself freely. He keeps watch over stereotypes and other prejudices and handles labeling with caution. Above all, the beginning professional is not desk-bound. He moves out into the community to help the people where they live. Although he may be comfortable with what he is now, he is uncomfortable if he must remain as he is.

Therefore he searches for ways to grow

through self-evaluation. In the process of looking at himself he finds gaps that he must fill with knowledge and refined skills. He does this through some form of continuing education. Furthermore, he looks deeper into self-evaluation of himself, of his competence with his clients, and of his relationships with others, and he discovers ways to measure his level of functioning. Accountability is this procedure of evaluation that describes the effectiveness of a person, a program, or an institution and also how it may become more effective. Accountability may be accomplished by forms, by scales, by questionnaires, by observers, and by records and reports. Self-disclosure is the basis for evaluation. However, evaluation is meaningless in a hostile atmosphere in which a punitive attitude of management strikes out at the employer. Positive progress depends on humanness.

—And all of this leads to each individual's somewhere. The somewhere of dignity and hope.

Appendix A

GLOSSARY

abnormality Term used with several meanings: (1) maladaptive behavior that is not in accordance with cultural requirements or environmental demands expected for age level, sex, or social role and (2) behavior away from or deviating from the norms (standards) of a particular society.

action/reaction interview Refers to a form of specialized conversation in which a one-directional, more or less impersonal form of communication between client and interviewer occurs.

actual self Includes all the behavioral and physical characteristics of an individual at any one moment, whether or not the individual or other people are aware of these characteristics.

adult ego state Term used in transactional analysis that refers to an individual's behavioral style (ego state), which is primarily logical, positive, and factual.

affective level Refers to the level or frequency of change of an individual's feelings or emotions.

ahistorical Approach to the study of people that concentrates on the here-and-now, present events.

alcologia Combination of the denial and rationalization typical of the alcoholic's reasoning.

alienation Process by which an individual detaches himself from others and hides within himself in a form of self-attachment.

ambivalent feelings Refer to simultaneous sensations of positive and negative emotions, for example, loving and hating the same person.

antecedent event (A) Refers to the stimulating or beginning events that happen before behavior (B).

anxiety hierarchy Refers to a reconditioning process in which a series of steps are planned in order to reduce the degree of anxiety associated with an object, person, or event.

appropriate behavior May be explained as the terminal behavior established by the behavior-changer (interviewer) and the client as the behavior to be strengthened or the coping mechanisms (learned behavior) that are tension-reducing for an individual and that result in a harmonious relationship with environmental events and with people around the individual.

assertive training Conditioning process during which an individual becomes better prepared to express himself in his relationships with people.

authenticity A characteristic of genuineness that refers to being oneself or being "real" and not a phoney.

automated interviewing Form of behaviorally oriented interview in which the computer is programmed to ask the client questions on printouts.

autonomy Self-direction and independence in decision making.

baseline data Refers to the frequency (number of times) of certain behavior within specified time periods.

behavior (B) May be explained as acts and feelings of an individual that may be seen and/or heard or concealed; what an individual says or does after the antecedent event (A).

behavior modification Process of changing behavior by means of reinforcement principles based on experimentally discovered laws of learning.

behavioral analysis Process of defining the target behavior to be eliminated and the terminal behavior to be strengthened.

belongingness Feeling of being part of, accepted by, and secure in relationships with one or more individuals.

blind-alley career Job that keeps worker locked into one kind of job and usually at one salary level.

body language Refers to the physical distance between people, the orientation or direction of the body (proxemics), and the motion of the body (kinesics).

brotherly love Nonsexual, unselfish caring for another human being.

case history Scientific biography of an individual, an institution, or group of people.

casual observation Nonscientific, unplanned "looking and listening" that is often biased and incomplete.

chaining Refers to a sequence of behaviors or a habit in which one conditioned behavioral response is the cue for the next response, for example, moving one foot after the other in order to walk.

child ego state Term in transactional analysis referring to an individual's behavioral style (ego state) that is likely to be impulsive, dependent, and open to parentlike demands of authority as well as carefree and joyful.

client-centered counseling Emphasizes the leading role of the client in decision making during the interaction between the client and the counselor.

clinical observation Organized form of naturalistic observation, for example, a case history.

closed-ended questioning Form of questioning that contains possible answers within the question, for example, multiple-choice questions.

code of ethics Refers to the rights of the client and the responsibilities of the helping person in maintaining the client's rights.

cognitive dissonance Process of receiving information that conflicts with an individual's beliefs and/or knowledge.

cognitive process Act or processes involved in thinking.

compensation Attempt to overcome feelings of inferiority or dissatisfaction with real or supposed weakness by developing strength in some other area.

conditioning Process of learning in which a new stimulus becomes associated with a former response (classical conditioning) or a new response is acquired when this response satisfies a need (operant conditioning).

confrontation Form of communication that calls attention to a person's behavior and challenges him to examine the effect of this behavior on himself and other people.

congruence A characteristic of genuineness that refers to behavior in which an individual's words and actions fit together.

consequent event (C) Event that follows behavior; the reinforcement that either strengthens or weakens the continuation of behavior.

contingencies Refer to behavior plus its consequences.

contingency management Conditioning process by means of which a relationship is established between behavior and the consequences or reinforcements of this behavior.

continuum Series of things, people, or events that have some relation to one another and differ in either degree or some quantity or quality.

controlled nondirection Form of interviewing in which the interviewer offers alternatives for problem solving and reinforces the client's self-selection.

coordinating/paraphrasing An expressive skill in which the listener organizes and re-words the speaker's message but maintains the speaker's sentiments.

counseling Form of interviewing that is concerned with self-exploration and/or behavioral examination.

coping The individual's behavioral responses to environmental demands and self-needs.

coping mechanisms Learned ways of behaving in the process of satisfying needs; acquired in order to curb the distress and/or tension resulting from environmental demands antagonistic to the satisfaction of needs. Coping mechanisms are similar to coping strategies and defense mechanisms.

cue Refers to the stimulus, the excitant, or the signal that elicits certain behavior.

daydreaming Process of thinking (imagining) while awake, which serves to satisfy unfulfilled needs (wishes).

demographic factors Statistical information about such items as the distribution of people in neighborhoods and their ages, races, sex, birthplaces, socioeconomic status, and education.

denial Form of behavior in which an individual avoids painful events or thoughts by pretending that the event does not exist.

dependent variable Behavior that follows the antecedent event.

depth interview Stresses and carefully explores the impact of underlying attitudes, feelings, and self-understanding on the client's behavior and problems.

descriptive words Those that explain what is

seen or heard in behavioral terms, for example, "walks" or "hits."

desensitization Conditioning process by means of which the level of anxiety associated with certain events (stimuli) is gradually reduced.

developmental approach Method of determining manpower needs that begins with exploration of needs and problems of the clients, their families, and their communities and then determines the tasks and activities that will satisfy these needs.

dialogue Form of communication in which both the message-sender and the message-receiver are listening and responding to one another.

didactic Refers to teaching that concentrates on theory, facts, and sometimes moral instruction.

direct questioning Information-seeking question that asks for a specific answer; for example, "What is your address?"

displacement Response to a situation considered threatening by an aggressive act against an object or a person other than the threatening source.

duelogue Communication in which two individuals arm themselves with word weapons and carry on a verbal duel.

duologue Similar to two monologues in which two individuals speak and try to control the conversation at the same or at separate times.

dyslexia Lowered ability to read and/or to understand what is read that is thought to be due to minimal brain dysfunction.

eductive technique Approach to an interview that encourages the client to talk about his concerns yet assures him that he is not obligated to do so.

ego-alien Rejection of a problem or characteristic that an individual feels does not fit the way he thinks of himself.

ego state Organized system of feelings and actions that is expressed as a parent, adult, or child behavioral style.

ego-syntonic Acceptance of a problem or characteristic that an individual feels fits into the way he thinks of himself.

emotional insulation Behavior that is indifferent, detached, and unemotional to protect the individual from emotional distress.

empathy Active, persistent, and respectful tuning in of one person to another's point of view to receive and share the other person's meaning and feelings about his experiences.

ethnic Refers to groups of people thought to be biologically and/or culturally related.

experiental interview Stresses the human relationship involved with two people "being"

together, sharing and changing thoughts and feelings in an acceptant atmosphere.

expressive feelings Free discussion of feelings (ventilating) that helps to relieve tension and usually gives the individual additional energy to achieve new understanding of his problem.

expressive skills Communications that attempt to accurately feed back the meaning and/or feeling of a message in order to bring new understanding.

external frame of reference Stresses viewpoints and/or observations that focus on behavior that can be seen or heard and objective observations made from the viewpoint of an outsider looking at the behavior of another person.

facial language Includes any movement or change in the face, for example, the wrinkled brow, the smile, or eye contact.

fantasizing Imaginative attempts that may result in a plan and action or a substitute for action.

feedback Refers to the return message from the message-receiver to the message-sender.

field observation Form of naturalistic observation that explores the behavior of animals and humans in their natural settings.

filial therapy Process of training parents with specific guidelines in the treatment of their disturbed children within a play group that includes other disturbed children.

fugue Individual's physical escape from his usual surroundings by more or less forgetting past identity and events.

functional professionals Individuals selected on the basis of their existing abilities in a given area who are trained to a high level of expertise in that area.

funnel questioning Forms of questions that begin with generalities and proceed to more specific questions.

generalist Refers to concern for the whole individual and his needs and also to training in a job that is related to working with people.

generalized behavior pattern Gradual expansion of certain newly learned ways of behaving to other similar situations.

generativity Interest in the production of offspring and/or in a parent role; also a creative process that includes caring for oneself and others with the goal of guiding the fulfillment of potentials.

genuineness Describes the person who is aware and acceptant of his own feelings and experiences and is able to freely and honestly communicate these feelings and experiences to someone else.

Gestalt observation Process of picking up cues about the observed person through all channels of communication and putting these all together for the pattern of behavior that emerges.

gestural language Movements of the body or parts of the body that substitute for or emphasize words.

Greenspoon effect Refers to sounds such as "mm-hmmm," "huh-uh," or other speech mannerisms as well as visual cues that reinforce the client to continue his responses in a certain direction.

"halo" effect Influence of the first impression of a person, event, or object that colors all later impressions either favorably or unfavorably.

"hello-goodbye" phenomenon Describes the behavior of the client who feels so much better after one or two interviews that he stops his interview visits, falsely believing his problems have been solved.

highlighting/paraphrasing An expressive skill in which the listener repeats several of the speaker's remarks that seem to be similar in meaning.

historical question Question that asks about past events.

hypotheses "Educated guesses," interpretations, inferences, and hunches to explain the relationship between various factors and human behavior or other events.

inappropriate behavior Coping mechanisms that result in behavior that is bothersome or destructive to the individual; the target behavior that has been established by the behavior-changer (interviewer and the client) as the behavior to be eliminated.

iatrogenic Refers to a disorder that is produced and sustained by the interviewer or some other person's suggestions.

identification Behaving like one's interpretation of the characteristics of significant people or objects; recognizing one's identity.

identity The unique ideas an individual has about his role and status in society that provides a sense of wholeness and of self-continuity over a period of time.

identity diffusion Inability to establish a feeling of wholeness and uniqueness of one's ideas, values, and feelings.

inclusion Feeling of being important, worthy, and part of a group; similar to belongingness.

incorporation Early form of recognizing external reality and of identification in which the infant and the young child take into themselves the behavioral qualities of another individual.

incubation period Time of silence during an interview when the client or the interviewer is putting together the pieces of his problem and/or his thoughts.

independent variable Antecedent event; the factor that is changed in an experiment to determine its effects on some form of behavior (dependent variable).

indigenous nonprofessional An individual living in the same slum area, belonging to the same income or ethnic group, and/or having experienced similar problems as the client.

indirect questions Explore for further communication and understanding by asking for information with broad, general statements; for example, "Tell me more about that situation."

information-seeking interview Directed to gathering data about people, places, or products.

information-sharing interview Goes one step beyond information seeking since the data collected is not solely for the interviewer's purposes but is also to be shared with the client.

insight game Persistent pursuit of understanding by means of questioning and discussion that is a cover-up for defending and continuing existing behavior; similar to intellectualization.

insight interview Focuses on the discovery of relationships among events, often of past experiences with present behavior; uses interpretation of the client's unconscious conflicts and behavioral responses as the basis for establishing the relationships; and concentrates on the client's unconscious processes as the originators of the client's symptoms.

intellectualization Defensive maneuver (coping strategy) that uses lengthy explanations and small details in order to conceal turbulent emotions and/or undesired thoughts.

intelligence May be defined according to the device that measures this ability—intelligence is what the intelligence test measures; qualitatively—intelligence is the increasing competence of an individual that develops in four main stages toward logical thinking (Piaget, 1952); and as a tool—a personality characteristic that serves in the process of adapting and/or problem solving.

interaction interview Specialized form of two-directional communication in which the interviewer listens to and responds to the client in order to assist the client in problem solving and decision making.

interdependence Feelings and behavior demonstrated in a cooperative and sharing approach in one's dealings with other people; making one's own decisions (independence), yet caring enough to help and be helped by others (dependence).

intermittent reinforcements Rewards and/or punishments given at preplanned intervals of time for the purpose of strengthening or weakening a particular behavior.

internal frame of reference A viewpoint that seeks understanding about another individual "from the inside looking out"; understanding gained through the examination of events from the viewpoint of the behaving person rather than from the viewpoint of the observer.

internalization Process of making certain ideas, standards, and behavior part of an individual's behavior pattern.

interpretation An expressive skill by which the counselor introduces his viewpoint in remarks about what the client is saying.

interpretive words Words that evaluate, pass judgment, and label behavior as "crazy," "dumb," "good," and so on.

intervention Refers to the helping professional's functions in remediation, prevention of maladaptive behavior, and the stimulation of positive development of individuals, groups, institutions, or communities through direct service, consultation and training, or communication media.

interviewing Specialized pattern of communication with defined goals and with emphasis on some degree of interpersonal relationship.

intimacy Refers to closeness plus acceptance of the worthiness of another person and also some degree of responsibility for the other person's welfare.

inverted funnel questioning Form of questioning that begins with requests for specific information and proceeds to more general areas.

introjection Process of adopting other people's values, attitudes, and behavior as one's own for the purpose of avoiding conflicts and in order to feel accepted by others.

job-factoring approach Focuses on an analysis of the specific tasks and related activities that various levels of workers perform.

Johari's window Diagram of the interpersonal behavior of an individual that reveals the degree of openness with which he meets other people and his level of awareness of himself.

journalistic interview Information-seeking interview conducted by a reporter to gather answers to the five W's: *Who? What? When? Where? Why?* and sometimes *How?*

kinesics Study of communication meanings of movements of the body and of the face.

leading questions Similar to closed questions that restrict answers to certain choices by including the expected answer within the question; for example, "You don't really want to go, do you?"

learning style Individual's preferred method of achieving understanding that is conditioned in early childhood by the important people around the child, for example, through hearing or through touching.

life space The "world" of an individual that includes his home, the people, and the geographical area in which he moves as well as the interaction of his needs and goals with these external factors.

life-style Behavior pattern of the individual that includes his appearance, values, attitudes, goals, and the procedures he uses to achieve his goals.

manipulative feelings Refer to emotional behavior used to urge or embarrass another individual to perform acts desired by the controlling individual.

mechanism General term for habits or behavioral acts by means of which individuals meet environmental demands while satisfying and protecting their needs.

mental retardation Impairment in adapting to natural and social demands of one's environment that may be shown in lower level of maturation of self-help skills such as sitting, crawling, walking, talking, habit training, and interaction with children; the inability to learn—to acquire knowledge from one's experiences; and lowered ability to maintain oneself independently and responsibly in the community and in gainful employment.

mental status examination Part of a psychiatric interview that takes into account emotional and intellectual functioning and primarily serves a diagnostic purpose.

microcounseling Brief and carefully planned series of steps with immediate feedback in which an individual is training in specific interviewing/counseling behaviors.

minimal encourages Procedure that encourages the client by means of respectful listening and well-timed remarks to carry on most of the conversation.

modeling In behavior change, refers to the training procedure in which the human service worker performs certain behavior that is imitated by the client.

monologue Communication process in which an individual takes over the conversation and expresses his thoughts out loud, disregarding the interests or responses of others.

motivation Refers to the external or internal stimulations for behavior.

multiple questions Contain two or more ideas; for example, "Would you like to stop here or go on?"

naturalistic observations On-the-spot running records of observations that begin and end at any point with no planned attempt to change the situation of the observed person.

negative reinforcements Unpleasant consequences of behavior with the goal of extinguishing inappropriate behavior.

nonjudgmental attitude Refers to behavior that does not place one's own values, estimates of worth, or criticisms on the opinions and/or behavior of the other person.

nonpantomimic gestures Actions that accompany words and modify or regulate the meaning of the words.

nonpossessive warmth Expression of commitment, concern, and respect for another human being as he is, not as he ought to be, and "with no strings attached."

nonprofessional Usually refers to nontraditionally trained workers without educational degrees who do not meet qualifications set by professional groups.

nonverbal cues Visual messages without sound that include body language, facial language, and gestures.

normality Refers to the usual, the average, or accepted behavior within a certain culture.

norms Values or criteria of behavior established for certain characteristics and/or for certain groups that become the standards by which normality is judged.

objective Behavior that is open to observation and is reported free from personal/emotional prejudices; goal to be achieved.

observation, scientific Cautiously planned, orderly process of fact-gathering through one's senses.

obsessive Refers to idea or impulse that persists even though the person prefers to be rid of it.

open-ended questioning Broad general request that invites the answerer to express his own views and feelings; for example, "How do you feel about this glossary?"

operant conditioning Process of strengthening or weakening certain voluntary and unprompted behavior by means of consequences (reinforcements).

pantomimic gestures Actions that substitute gestures for words.

paraphrasing Form of translation in which the listener expresses the sense of what he has just heard so that the chief points are pulled together.

paraprofessional Individual functioning in a similar but secondary role that resembles the activities of the professional but carries less responsibility and less decision making.

parent ego state Organized system of feelings and behavior that arises from experiences with parental figures and in which the behavioral style is judgmental, moralistic, and caring about the other person.

participant observation Careful "looking at and listening to" the observed person while becoming involved in the activities of the observed person.

perception Sensory impressions (seeing, hearing, smelling, feeling, and touching) modified and given meaning by life experiences.

personnel interview Information-seeking, sometimes information-sharing, interview conducted between an employee and an interviewer for the purpose of job placement and resolution of job or personal problems.

phobia (phobic) Intense, unreasonable fear that goes beyond the seriousness and/or danger of the object or situation.

positive reinforcements Pleasant consequences of behavior with the goal of strengthening appropriate behavior.

privileged communication Stresses that conversation between client and interviewer must be considered private and confidential.

process limits The degree and manner in which the client and the interviewer are expected to participate in interview situation.

professional Describes an individual who is knowledgeable about his particular area, skilled in procedures to be used in this area, functioning at a level which uses his skills for the client's benefit, and who maintains a code of ethics.

professionalitis Describes the interviewer who may feel uncomfortable in his interviewer role and hides behind complicated explanations and high-sounding words to hide his uncertainty and discomfort.

professional self Behavior resulting from training in knowledge and skills in order to satisfy established qualifications for certain levels of responsibility, competence, and also a code of ethics.

projection Refers to the shifting of one's unacceptable thoughts, feelings, and other undesired qualities onto someone else so that the blame for those thoughts or feelings is apparently transferred to the other person.

prompting Interviewer's verbal or nonverbal behavior that encourages the client to continue a certain behavior (discussion or action).

proxemics Study of body orientation and direction.

psychoanalysis Viewpoint about behavior of individuals that focuses on unconscious processes and the inner conflicts that initiate and maintain the client's symptoms.

psychodynamic forces Those interpersonal (between people) and intrapersonal (within a person) ideas, impulses, emotions, thoughts, and other characteristics that develop, influence, and change interview participants.

psychoecology Focuses on the interdependent influence of people, animals, things, and physical surroundings on an individual's behavior.

psychological tests Standardized (experimentally constructed and scaled) measures of characteristics such as interests, attitudes, achievements, and intelligence.

psychotherapy Specialized pattern of verbal and nonverbal interaction with specific purposes that goes deeper into self-exploration and self-disclosure than counseling and in which basic personality reorganization is usually one of the goals.

public opinion interview Information-seeking interview that surveys the opinions and/or attitudes of people toward some event or person.

rapport Refers to the harmonious relationship established in an interpersonal transaction.

rational Process of reasoning and logical problem solving.

rationalizing The effort to appear reasonable in order to cover up feelings, thoughts, or acts.

rational-emotive therapy Concentrates on the origin and development of emotional disturbance from irrational and illogical thoughts and philosophies.

reaction formation Shifting of unacceptable thoughts, feelings, and behavior into their more socially acceptable opposites, for example, masking a feeling of hostility with behavior expressing love and/or sympathy.

reciprocal affect Process by means of which one individual expresses a certain emotion (affect) that calls forth similar emotional behavior in the other person.

reciprocal inhibition Gradual reduction of anxiety feelings about a person, idea, or event by means of the introduction of behavior antagonistic to the anxiety-provoking stimuli that suppresses and eventually weakens the anxiety reaction, for example, relaxation when presented with the feared object.

reflection Act of uncovering and making known in fresh words the feelings that lie within the client's comments.

reinforcement Concerned with the positive and negative consequences (rewards and punishments) that either strengthen and continue or weaken and discontinue certain behavior.

reinforcement menu Refers to the list of negative consequences (punishments) to be used in eliminating one kind of behavior (target behavior) and the positive consequences (rewards) for the appropriate behavior (terminal behavior).

repeating/paraphrasing Feedback of one or more words in order to direct the speaker into further discussion of a certain topic and to assure the speaker he is being understood and attended to carefully.

regression Less mature behavior that is representative of an earlier developmental level.

repression Unaware (unconscious) holding back or covering up of ideas, feelings, or acts that an individual considers too painful to himself or to others or that he feels would be socially unacceptable.

response repertoire Refers to a collection of skills and knowledge that an individual learns and performs comfortably and effectively.

right to treatment Client's right to demand help and to receive adequate treatment, to refuse a particular kind of treatment, and to seek another kind of help elsewhere.

role reversal Occurs when one individual changes his behavior and acts as if he were the other person.

role-playing Training method in which individuals test and receive immediate feedback about approaches to a situation by acting out the handling of a situation and/or by trying out new skills.

roles Individually interpreted patterns of culturally expected behavior.

script analysis Term used in transactional analysis referring to the exploration and careful examination of the individual's life plan that serves as the basis (motivation) for the individual's behavior.

secondary source observation Information about the observed person obtained from files, reports, or other people.

self-actualization Process of recognizing and accepting oneself, developing abilities and satisfying interests, and establishing a harmonious relationship between individual expression and social interest.

self-assertion Act of responding rather than holding back in social situations that may be expressed either with respect for the goals of others (assertion) or disregarding the goals of others (aggression).

self-concept An individual's values, attitudes, abilities, and behavior that he recognizes as his own and evaluates favorably or unfavorably.

self-fulfilling prophecy Refers to an individual's expectations regarding a situation; because the individual believes certain results inevitable, he acts in such a way that he influences the expectations to actually happen, thus strengthening his belief in the self-fulfilling prophecy, for example, acting aggressively because he expects the other person to be aggressive.

self-fulfillment Similar to self-actualization with the possible difference that self-fulfillment focuses on satisfying an individual's needs while self-actualization is also concerned with the needs of others.

shaping Procedure in which terminal behavior is broken into small units or single responses that are conditioned and each small unit of behavior builds on (chaining) behavior already learned (conditioned).

social system Refers to the effect on behavior of variations in the interaction of two or more persons as well as the environmental factors around these individuals.

soliloquy Act of speaking to oneself out loud.

spectator observation Careful "looking and listening" within the same room but not near the observed person.

standardized observation Carefully planned and controlled "looking and listening."

standards Expected levels of performance that are determined by the individual himself, by other people, or by institutions.

status Refers to culturally and individually determined positions stemming from sex, age, race, and education that determine the rights and obligations of a person.

stimulus situation Refers to a pattern of antecedent events.

structural analysis Term used in transactional analysis that refers to the progressive analysis of the organization and interaction of ego states.

subprofessional Describes an individual whose responsibilities are less complicated than the professional and whose functioning is usually expected to be subordinate to the professional.

summarization An expressive skill in which there is a restatement of what the speaker has been saying or feeling during an entire counseling session or a series of counseling sessions.

suppression Form of voluntary forgetting that is usually temporary.

survey interview Information-seeking interview that seeks data about attitudes, for example, preferences for certain boxes for a new food product.

sympathism Sympathy-seeking behavior that asks for emotional support by dwelling on misfortunes and/or physical complaints; openness to the influence of someone's suggestions.

sympathy Many forms of expression that range from pity to compassion; differs from empathy since it signifies that the sympathetic individual shares the same feelings as the other individual and it sometimes implies a value judgment that the sympathetic person is more capable or "better off" than the individual who is pitied.

thought-stopping Program of conditioning in which annoying thoughts are gradually weakened and finally eliminated by means of negative reinforcement such as saying "Stop" whenever the disturbing thoughts arise.

time interval or time-sampling Requires observations at predetermined times of the day or night and usually for a set length of time.

transaction contract Mutually designed written or oral listing of the client's and the interviewer's promises and objectives.

transaction interview Social system of shared responsibilities in which client and interviewer establish rules for self-discovery and personal growth and in which the relationships that result influence both client and interviewer to change.

transactional analysis Focuses on the discovery and script analysis of the predominant ego state of parent, adult, or child and the influence of the behavioral style of this ego state in helping or hindering the individual in his encounters with people and other events.

triangulation Refers to the pulling in of the interviewer or some other person as an ally against someone whom the client finds disturbing.

Truax triad Refers to the three characteristics, empathy, genuineness, and nonpossessive warmth, that are considered essential to the effective counseling situation.

undoing Way of escaping painful events or thoughts by constantly atoning for real or imaginary sins.

ventriloquizing Refers to an individual who speaks as if he were commenting for someone else; for example, "Lots of people are asking about how you feel about this book."

withdrawal Pattern of behavior in which an individual removes himself physically or psychologically from disturbing circumstances.

Appendix B

SELECTED PSYCHOLOGICAL TESTS FOR SELF-UNDERSTANDING

The Adjective Check List, H. G. Gough and A. B. Heilbrun, Consulting Psychologists Press, 577 College Avenue, Palo Alto, Calif. 94304. (Gives 300 adjectives commonly used to describe attributes of a person; may be used for self-evaluation or be rated by someone other than subject.)

Caring Relationship Inventory, E. L. Shostrom, Educational and Industrial Testing Service, San Diego, Calif. 92107. (Measures the essential elements of caring or love.)

IPAT Anxiety Scale Questionnaire, R. B. Cattell, Institute for Personality and Ability Testing, Champaign, Ill. 61820. (Arrives at six scores: self-sentiment development, ego strength, protension of paranoid trend, guilt proneness, ergic tension, and total anxiety.)

Personal Orientation Inventory for the Measurement of Self-Actualization, E. L. Shostrom, Educational and Industrial Testing Service San Diego, Calif. 92107. (Emphasizes mentally healthy and actualizing qualities rather than pathological characteristics. Growth toward self-actualization may be assessed by administering this to trainees at the beginning and toward the end of training. May be used for similar purpose with clients.)

Sixteen Personality Factor Questionnaire (16 PF Test), R. R. Cattell, Institute for Personality and Ability Testing, Champaign, Ill. 61820. (Measures 16 main dimensions of personality. Has two forms each with 187 items [forms A and B] and one shorter form [C] with 105 items.)

Standards for Educational and Psychological Tests and Manuals, American Psychological Association, 1200 Seventeenth Street, N.W., Washington, D. C. 20036.

Appendix C

SOURCES FOR AUDIOVISUAL AND OTHER AIDS FOR HUMAN SERVICE CURRICULUM AND HUMAN SERVICE WORKER

ACCOUNTABILITY AND EVALUATION

Evaluation: a Forum for Human Service Decision-makers, Program Evaluation Project, 501 Park Avenue South, Minneapolis, Minn. 55415. (Communication medium for persons interested in the evaluation of human services.)

Nine Evaluation Aids from CSE, Center for the Study of Evaluation, 145 Moore Hall, University of California, Los Angeles, Calif. 90024. (A list of aids for educational evaluation and curriculum improvement.)

BIBLIOGRAPHIES

Early Childhood Psychosis: Annotated Bibliography, 1964-1969, C. Q. Bryson and J. H. Hingtgen, National Clearinghouse for Mental Health Information, Superintendent of Documents, Washington, D. C. 20402. ($1.25.)

Educators' Guide to Free Guidance Materials, a Multimedia Guide, ed. 12, 1973, Educators Progress Service, Randolph, Wis. 53956. (Published annually.)

Guide Book Describing Pamphlets, Posters, Films on Health and Disease, ed. 3, 1967, Office of Public Health Education, Maryland State Department of Health, 301 West Preston Street, Baltimore, Md. 31201.

Guide to the Literature in Psychiatry, 1971, B. Ennis, Partridge Press, Los Angeles. (Comprehensive list of journals, publications, books, government documents, directories, and sources for audiovisual aids.)

Human Intelligence, L. Wright, National Clearinghouse for Mental Health Information, Superintendent of Documents, Washington, D. C. 20402. ($2.50.)

Inservice Training for Allied Professionals and Nonprofessionals in Community Mental Health, Public Health Service Publication No. 1901, Superintendent of Documents, Washington, D. C. 20402.

Inservice Training for Key Professionals in Community Mental Health, Public Health Service Publication No. 1900, Superintendent of Documents, Washington, D. C. 20402.

Medical Model Bibliography, Department of Psychology, University of Windsor, Ontario, Canada.

Multi Media Store Catalog, American Personnel and Guidance Association, 1607 New Hampshire Avenue N.W., Washington, D. C. 20009. (A list of periodicals, books, pamphlets, films, and cassette tapes.)

National Association for Retarded Children, 2709 Avenue E East, Arlington, Texas 76011. (Publications list.)

New Careers Bibliography: Paraprofessional in the Human Services, 1970, C. R. Vestal and S. K. Craig, National Institute for New Careers, National Technical Information Service, U. S. Department of Commerce, 20402.

Pornography: Review and Bibliographic Annotations, J. Money and R. Athanasion, Am. J. Obstet. Gynecol. **115**:130-146, 1973. (Reprint.)

The Psychoanalytic Study of the Child, C. L. Rothgeb, S. M. Clemens, and E. M. Lloyd, National Clearinghouse for Mental Health Information, Superintendent of Documents, Washington, D. C. 20402. (Abstracts.)

Science for Society, American Association for the Advancement of Science, National Science Foundation, Washington, D. C. 20550.

Self-Help Materials, developed by T. J. O'Farrell, P.O. Box 185, Williamsburg, Maine 02113.

Special Learning Materials for the Special Child and Those with Learning Difficulties, Follett Publishing Co., Chicago, Ill.

Suicide and Suicide Prevention, N. L. Farberow, Department of Health, Education, and Welfare Publication, Superintendent of Documents, Washington, D. C. 20402. ($2.00.)

Superintendent of Documents, U. S. Government Printing Office, Washington, D. C. 20402. (Send for a free list containing government publications.)

Training Methodology, Public Health Service Publication, Superintendent of Documents, Washington, D. C. 20402. (Separate annotated bibliographies on background and research, Part I; planning and administration, Part II; and audiovisual theory, aids, and equipment, Part IV.)

A Training Resource Guide for Human Services, D. R. Slavin and W. J. Gordon, Southern Regional Education Board, 130 Sixth Street N. W., Atlanta, Ga. 30313.

Understanding Body Movement, 1973, M. Davis, Arno Press, Inc., New York, N. Y. (An annotated bibliography.)

NEWSLETTERS

Aclohol and Health Notes, National Institute on Alcohol Abuse and Alcoholism, National Clearinghouse for Alcohol Information, Box 2345, Rockville, Md. 20852. (Published monthly.)

Allied Health Trends, Association of Schools of Allied Health Professions, One Dupont Circle, Suite 300, Washington, D. C. 20036.

Behavior Today, P.O. Box 2993, Boulder, Colo. 80302. (Published weekly; $25.00 per year.)

ERIC, Clearinghouse on Educational Media and Technology, Institute for Communication Research, Stanford University, Stanford, Calif. 94305.

Gay People and Mental Health, Box 3592, Upper Nicollet Station, Minneapolis, Minn. 55403. (Monthly bulletin for purpose of sharing information about counseling homosexuals, community education, and sex education; $6.00 per year.)

Health News, State of California Health and Welfare Agency, Department of Health, 714 P Street, Sacramento, Calif. 95814. (Published "for health professionals and those concerned with health issues in California, and, also, has relevance for human service workers outside of California.")

IRCD Bulletin, Horace Mann-Lincoln Institute, Teachers College, Columbia University, 525 West 120th Street, New York, N. Y. 10027. (Publication of the ERIC Information Retrieval Center on the Disadvantaged.)

Karuna, Center for Human Services Research,

Johns Hopkins University, Room 212, Harriet Lane, Obs. 2, Johns Hopkins Hospital, 601 N. Broadway, Baltimore, Md. 21205.

Keeping Up, Clearinghouse for Social Studies/Social Science Education, 855 Broadway, Boulder, Colo. 80302.

Liaison, 744 P Street, Suite 724, Sacramento, Calif. 95814. (Lanterman Mental Retardation Services Act, July, 1971, State of California; monthly publication.)

Mental Hygiene News, New York State Department of Mental Hygiene, 44 Holland Avenue, Albany, N. Y. 12208.

Periodically, American Psychological Association's Clearinghouse on Precollege Psychology and Behavioral Science, 1200 Seventeenth St. N.W., Washington, D. C. 20036. (Information on teaching psychology at the secondary school and elementary school levels with reviews, articles, and gimmicks for classroom use; published monthly September through May.)

Report on Education Research, Capital Publications, Inc., Suite G-12, 2430 Pennsylvania Avenue N.W., Washington, D. C. 20037. (Biweekly news service devoted to basic and applied research in education.)

Resources for Youth Newsletter, The National Commission on Resources for Youth, 36 West 44th Street, New York, N. Y. 10036. (Information on innovative programs that provide youth with opportunities for rewarding and responsible roles in society.)

Special Report of the President's Committee on Employment of the Handicapped, Superintendent of Documents, Washington, D. C. 20210. (Contains brief descriptions of programs for employment and care of the disabled.)

STASH Capsules, The Student Association for the Study of Hallucinogens, Inc., 638 Pleasant Street, Beloit, Wis. 53511. (Forthright and scientific account of nonmedically prescribed drugs in use.)

The Exchange, 311 Cedar Avenue South, Minneapolis, Minn. 55404. (Monthly newsletter for hotline, switchboard, and related services; $6.00 per year.)

RESOURCES FOR INFORMATION ABOUT HUMAN SERVICES

Career Development, Human Services Press, University Research Corporation, Suite 1600, 5530 Wisconsin Avenue N.W., Washington, D. C. 20015. (Contains reports on health manpower, new careers, books, and films; $10.00 per year.)

The Center for Human Services Research, Room 212, Harriet Lane, Obs. 2, The Johns Hopkins Hospital, 601 N. Broadway, Baltimore, Md. 21205.

Citizen's Advocacy for the Handicapped, Impaired and Disadvantaged: an Overview, Special Report of The President's Committee on Mental Retardation, W. Wolfensberger, Superintendent of Documents, Washington, D. C. 20402.

Directory of Facilities for the Learning-Disabled and Handicapped, 1972, C. Ellingson and J. Cass, Harper & Row, Publishers, New York, N. Y.

Drug Abuse Treatment Programs National Directory, 1972, D. D. Watson, National Clearinghouse for Drug Abuse Information, Superintendent of Documents, Washington, D. C. 20402.

Ethical Principles in the Conduct of Research With Human Participants, American Psychological Association, 1200 Seventeenth Street N.W., Washington, D. C. 20036.

Mental Health Associates of Maryland, Ms. Rhoda Levin, Essex Community College, Baltimore, Md. (Group organized by graduates of 2-year associate arts programs [human services, mental health technology, mental health associates] to investigate job market for graduates, to strengthen MHA identity, and to plan continuing education programs.)

Mental Health Directory, 1971, National Clearinghouse for Mental Health Information, Superintendent of Documents, Washington, D. C. 20402. ($3.75.)

National Clearinghouse for Drug Abuse Information, 5600 Fishers Lane, Rockville, Md. 20852. (Provides information on request through publications and a computerized information service; distributes publications.)

National Directory of Hotlines and Youth Crisis Centers, 1973, The National Exchange, 311 Cedar Avenue South, Minneapolis, Minn. 55404. ($2.00 per copy.)

Occupational Mental Health, Center for Occupational Mental Health, The Westchester Division of New York Hospital, Cornell University Medical Center, N. Y. (Contains article and abstracts related to mental health in occupations; published quarterly.)

Occupational Outlook Quarterly, Superintendent of Documents, Washington, D. C. 20402. (Contains reports concerning present and projected manpower needs and changing careers; published quarterly; $1.50 per year.)

Perspectives in the Field of Mental Health, a View of the National Program, 1972, R. H. Williams, Superintendent of Documents, Washington, D. C. 20402.

Schizophrenia Bulletin, Center for Studies of Schizophrenia and National Clearinghouse for Mental Health Information, Superintendent of Documents, Washington, D. C. 20402.

Social and Rehabilitation Record, Superintendent of Documents, Washington, D. C. 20402. (Information about new developments in habilitation of mentally retarded persons and rehabilitation of individuals with developmental problems; quarterly publication; $6.40 per year.)

Southern Regional Education Board, 130 Sixth Street N.W., Atlanta, Ga. 30313.

Western Interstate Commission for Higher Education, P.O. Drawer P, Boulder, Colo. 80302.

Women Studies Abstracts, P.O. Box 1, Ruch, N. Y. 14543. (Quarterly publication; $7.50 per year.)

SIMULATION GAMES AND RATING SCALES

A Drug Education Simulation Game, Public Document Distribution Center, 5801 Tabor Avenue, Philadelphia, Pa. 19120. (Two- to five-hour game that is useful for teachers inservice, student, and community groups. Through role-playing the participants attempt to define the nature and extent of the problem and to determine ways to deal with the problem. Cost of kit: $13.75.)

Ethical Judgment Scale, 1972, W. H. Van Hoose and C. F. Goldman, Wayne State University, Detroit, Mich. 48202. (Contains fifteen incidents with counselor and client with multiple choices of counselor replies to specified problem.)

Ethics in Counseling: Problem Situations, 1972, H. D. Christiansen, University of Arizona Press, Box 3398, Tucson, Ariz. 85722. (Each problem involves four hypothetical counselors representing different points of view.)

The Self-Disclosure Questionnaire. In *The Transparent Self*, 1971, S. M. Jourard, D. Van Nostrand Co., New York, N. Y. (Sixty-item questionnaire classified into six categories with ten questions in each category. Directed to discovering the degree to which an individual discusses certain items with parents, friends, and spouse.)

Simulation/Gaming/News, Box 8899, Stanford University, Stanford, Calif. 94305. (Bimonthly publication.)

TAPES, CASSETTES, RECORDS, FILMS, AND WORKBOOKS

The Art of Helping, 1973, R. R. Carkhuff, Human Resources Development Press, Box 222, Amherst, Mass. 01002. (Guide for developing skills in attending, responding, initiating, and communicating; $4.95.)

The Art of Problem-Solving, 1973, R. R. Carkhuff, Human Resources Development Press, Box 222, Amherst, Mass. 01002. (Guide for developing problem-solving skills for parents, teachers, counselors, and administrators; $4.95.)

Black Ivory, W. Robinson, Ampex, Perception Records Inc., Elk Grove Village, Ill. 60007. (Cassette that presents the black experience and is excellent for observation of verbal cues.)

The Case of Jim, J. Seeman, American Guidance Service, 2106 Pierce Avenue, Nashville, Tenn. 37212. (An annotated script and a record of counseling a client who stutters.)

Counseling: Today and Tomorrow, The American Personnel and Guidance Association, 1607 New Hampshire Avenue N.W., Washington, D. C. 20009. (Cassette series.)

Crisis Intervention Resource Manual, 1973, P. Mills, editor, School of Education, University of South Dakota, Vermillon, S. D. 57069. (Discusses training program for crisis intervention services and methods for developing skills in listening and interpersonal approaches; $4.00.)

The Eye of the Beholder, Stuart Reynolds Productions, 9465 Wilshire Blvd., Beverly Hills, Calif. 90212. (Excellent film to show the influence of perception on what an individual sees and hears.)

A Fuzzy Tale, 1969, C. M. Steiner, 135 Westminister, Kensington, Calif. 94708. (A fantasy that leads to a discussion of empathy.)

Helping Skills: A Basic Training Program, 1973, S. J. Danish and A. L. Hauer, Behavioral Publications, Inc., New York, N. Y. (Student's workbook and leader's manual.)

Instructional Strategies and Curricula for Secondary Behavioral Sciences, 1973, J. M. Johnson, editor, State University of New York, College of Arts and Science, Plattsburg, N. Y. 12901. (Contains curriculum units, classroom aspects, and teaching strategies.)

Learn Deep Muscle Relaxation, J. N. Marquis, Self-Management Schools, 745 Dishel Drive, Los Altos, Calif. 94022. (Instruction program on mimeographed sheets and cassette or 5-inch reel; $9.95.)

Listen In, Scott, Foresman & Co., Glenview, Ill. (Six diagnostic interviews with individuals or families seeking professional help.)

Psychoanalysis and Unconscious Racism, H. F. Butts, Behavioral Sciences Tape Library, 465 Main Street, Fort Lee, N. J. 07024. (Cassette.)

Other material may be obtained from the following groups:

Behavioral Sciences Tape Library, 485 Main Street, Fort Lee, N. J. 07024. (Has tapes on subjects such as child and adolescent psychotherapy, clinical and social psychiatry, learning, disabilities, family and group therapies, and feminism.)

The Center for Cassette Studies, Inc., 8110 Webb Avenue, North Hollywood, Calif. 91605. (Has cassettes on subjects relating to psychology, education, sociology, contemporary world affairs, arts, scientific affairs, and music.)

The Center for the Study of Democratic Institutions, Box 4446, Santa Barbara, Calif. 93103. (Publishes articles, a periodical, and tapes.)

The Human Development Institute, International Dynamics, Inc., 166 East Superior Street, Chicago, Ill. 60611. (Tape recordings.)

Indiana University Audio-Visual Center, Bloomington, Ind. 47401. (Films related to human services.)

Learning Dynamics, Inc., Eliot Building, 167 Corey Road, Boston, Mass. 02146. (Tape recordings.)

Mass Media Ministries, 2116 N. Charles Street, Baltimore, Md. 21218. (Films about human development and interpersonal relations.)

McGraw-Hill Films, 1221 Avenue of the Americas, New York, N. Y. 10020. (Films related to human services.)

McGraw-Hill Sound Seminar Series, P.O. Box 402, Hightstown, N. J. 08520. (Tape recordings.)

Media R/A, Inc., P.O. Box 2067, Van Nuys, Calif. 91404. (Tape recordings.)

National Education Tape Foundation, 2775 Cottage Way, Suite 31, Sacramento, Calif. 95825. (Tape recordings made at national and regional conferences by recognized authorities in psychology and education.)

National Medical Audiovisual Center, Atlanta, Ga. 30333. (Films related to human behavior and human services.)

New York University Film Library, 26 Washington Place, New York, N. Y. 10003.

Olympic Film Service, 161 West 22nd Street, New York, N. Y. 10011. (Publishes film lists.)

Psychological Films, 205 West Twentieth Street, Santa Ana, Calif. 92706. (Films about actualization, different forms of psychotherapy, and family therapy.)

Research Press, Box 31770, Champaign, Ill. 61820. (Publications and films concerning personal relationships.)

References

Ackerman, N. W., editor: Family process, New York, 1970, Basic Books, Inc., Publishers.

Adams, H. J.: The progressive heritage of guidance: a view from the left, Personn. Guid. J. 51:531-537, 1973.

Adler, A.: Social interest: A challenge to mankind (translated by J. Linton and R. Vaughan), New York, 1964, G. P. Putnam's Sons. (Paperback edition by Capricorn Books.)

Adler, G., and Myerson, P. G.: Confrontation in psychotherapy, New York, 1973, Science House, Inc.

Albee, G. W.: Mental health manpower trends, New York, 1959, Basic Books, Inc., Publishers.

Alberti, R. E., and Emmons, M. L.: Your perfect right, San Luis Obispo, Calif., 1970, Impact.

Allen, V. L.: Psychological factors in poverty, Chicago, 1970, Markham Publishing Co.

Allport, G. W.: Becoming: basic considerations for a psychology of personality, New Haven, 1955, Yale University Press.

American Psychological Association: Casebook on ethical standards of psychologists, Washington, D. C., 1967, The Association.

Angyal, A.: Neurosis and treatment: a holistic theory, New York, 1965, John Wiley & Sons, Inc.

Ansbacher, N. L., and Ansbacher, R. R.: Alfred Adler, superiority and social interest, Evanston, Ill., 1964, Northwestern University Press.

Aquilera, D. C., and Messick, J. M.: Crisis intervention: theory and methodology, St. Louis, 1974, The C. V. Mosby Co.

Arbuckle, D. S.: The counselor: who? what? Personn. Guid. J. 50:785-790, 1972.

Arbuckle, D. S., editor: Counseling and psychotherapy: an overview, New York, 1967, McGraw-Hill Book Co.

Ard, B. N., Jr., editor: Counseling and psychotherapy, Palo Alto, Calif., 1966, Science & Behavior Books, Inc.

Arnhoff, F. N., Rubenstein, E. A., and Speisman, J. C., editors: Manpower for mental health, Chicago, 1969, Aldine Publishing Co.

Aspy, D. N.: Empathy—congruence—caring are not singular, Personn. Guid. J. 40:637-640, 1970.

Avila, D. L., Combs, A. W., and Purkey, W. W.: The helping relationship sourcebook, Boston, 1971, Allyn & Bacon, Inc.

Baker, E. J.: The mental health associate: a new approach to mental health, Community Ment. Health J. 8:281-291, 1972.

Banaka, W. H.: Training in depth interviewing, New York, 1971, Harper & Row, Publishers.

Bandura, A.: Principles of behavior modification, New York, 1969, Holt, Rinehart & Winston, Inc.

Banks, G.: Black confronts white: the issue of support in the interview situation. In Carkhuff, R. R.: The development of human resources, New York, 1971, Holt, Rinehart & Winston, Inc.

Banks, W., and Martens, K.: Counseling: the reactionary profession, Personn. Guid. J. 51:457-462, 1973.

Baratz, S. S.: Effect of race of experimenter, instructions, and comparison population upon level of reported anxiety in Negro subjects, J. Pers. Soc. Psychol. 7:194-196, 1967.

Barbara, D. A.: The art of listening, Springfield, Ill., 1958, Charles C Thomas, Publisher.

Barclay, J. R.: Foundations of counseling strategies, New York, 1971, John Wiley & Sons, Inc.

Barker, R. G.: Ecological psychology, Stanford, Calif., 1968, Stanford University Press.

Barrett-Lennard, G. T.: Dimensions of therapists as causal factors in therapeutic change, Psychol. Monogr. **43**:76, 1962.

Beier, E. G.: The silent language of psychotherapy, Chicago, 1966, Aldine Publishing Co.

Bellak, L., and Loeb, L., editors: The schizophrenic syndrome, New York, 1969, Grune & Stratton, Inc.

Benjamin, A.: The helping interview, Boston, 1969, Houghton Mifflin Co.

Berenson, B. G., and Carkhuff, R. R., editors: Sources of gain in counseling and psychotherapy, New York, 1967, Holt, Rinehart & Winston, Inc.

Berne, E.: Principles of group treatment, New York, 1966, Grove Press, Inc.

Berne, E.: Games people play, New York, 1964, Grove Press, Inc.

Berne, E.: Transactional analysis in psychotherapy, New York, 1961, Grove Press, Inc.

Bertalanffy, L. von: General systems theory, New York, 1968, George Braziller, Inc.

Bird, C.: Born female, rev. ed., New York, 1971, Pocket Books.

Birdwhistell, R. L.: Kinesics and context: essays on body motion, Philadelphia, 1970, University of Pennsylvania Press.

Blank, L., Gottesgen, G., and Gottesgen, M., editors: Encounter: confrontations in self and interpersonal awareness, New York, 1971, The Macmillan Co.

Brammer, L. M.: The helping relationship, Englewood Cliffs, N. J., 1973, Prentice-Hall, Inc.

Brammer, L. M., and Shostrom, E. L.: Therapeutic psychology: fundamentals of actualization counseling and psychotherapy, Englewood Cliffs, N. J., 1968, Prentice-Hall, Inc.

Brigham, J. C., and Weissbach, T. A., editors: Racial attitudes in America: analysis and findings of social psychology, New York, 1972, Harper & Row, Publishers.

Brotman, R., and Freedman, A.: A community mental health approach to drug addiction, Washington, D. C., 1968, U. S. Department of Health, Education, & Welfare.

Brown, P. B., editor: Radical psychology, New York, 1973, Harper & Row, Publishers.

Buber, M.: I and thou, New York, 1958, Charles Scribner's Sons.

Bugental, J. F. T.: The search for authenticity: an existential approach to psychotherapy, New York, 1965, Holt, Rinehart & Winston, Inc.

Burton, A.: Interpersonal psychotherapy, Englewood Cliffs, N. J., 1972, Prentice-Hall, Inc.

Butler, R. N., and Lewis, M. I.: Aging and mental health, St. Louis, Mo., 1973, The C. V. Mosby Co.

Butts, H. F.: Psychoanalysis and unconscious racism, Fort Lee, N. J., 1972, Behavioral Sciences Tape Library. (Cassette.)

Carkhuff, R. R.: Credo of a militant humanist, Personn. Guid. J. **51**:237-242, 1972.

Carkhuff, R. R.: Development of human resources, New York, 1971, Holt, Rinehart & Winston, Inc.

Carkhuff, R. R.: Helping and human relations: a primer for lay and professional helpers, New York, 1969, Holt, Rinehart & Winston, Inc., vols. I and II.

Carkhuff, R. R., and Berenson, B. G.: Beyond counseling and therapy, New York, 1967, Holt, Rinehart & Winston, Inc.

Carnes, G. D.: Identifying clients predisposed to failure, J. Counsel. Psychol. **20**:79-83, 1973.

Carney, R. E., editor: Risk-taking behavior, Springfield, Ill., 1971, Charles C Thomas, Publisher.

Casriel, D. A.: A scream away from happiness, New York, 1972, Grosset & Dunlap, Inc.

Chiang, H., and Maslow, A. H., editors: The healthy personality, New York, 1969, Van Nostrand Reinhold Co.

Cohen, D. H., and Stern, V.: Observing and recording the behavior of young children, New York, 1970, Teachers College Press.

Cohen, W. J.: Revolution in mental health, Personn. Administr. **32**(2):4-8, 1969.

Colbert, J., and Hohn, M.: Guide to manpower training, New York, 1971, Behavioral Publications, Inc.

Colebrook, J.: The cross of lassitude: portraits of five delinquents, New York, 1967, Alfred A. Knopf, Inc.

Combs, A. W., and Snygg, D.: Individual behavior: a perceptual approach to behavior, New York, 1959, Harper & Row, Publishers.

Combs, A. W., et al.: Florida studies in the helping professions, Monograph 37, Gainesville, Fla., 1969, University of Florida.

Coopersmith, S.: The antecedents of self-esteem, Berkeley, Calif., 1967, University of California Press.

Corsini, R. J.: Roleplaying in psychotherapy: a manual, Chicago, 1966, Aldine Publishing Co.

Counseling and the social revolution, Personn. Guid. J. **49**:special issue, 1971.

Cowen, E. L., Gardner, A., and Zax, M.: Emergent approaches to mental health problems, New York, 1967, Appleton-Century-Crofts.

Curran, C. A.: Counseling-learning: a wholeperson model for education, New York, 1972, Grune & Stratton, Inc.

Dahms, A. M.: Emotional intimacy, Boulder, Colo., 1972, Pruett Publishing Co.

D'Amico, R., and Todd, S. P.: Financing human service programs, Minneapolis, Minn., 1970, Institute for Interdisciplinary Studies, American Rehabilitation Foundation.

Deutsch, M.: The disadvantaged child and the learning process. In Reissman, F., Cohen, J., and Pearl, A.: Mental health of the poor, New York, 1964, The Free Press.

Dilley, J. S.: Anti-shrinkthink, Personn. Guid. J. **50:**567-572, 1972.

Dinkmeyer, D. C., and Moro, J. J.: Group counseling: theory and practice, Itasca, Ill., 1971, F. E. Peacock Publishers, Inc.

Duncan, D. D., Featherman, D. L., and Duncan, B.: Socioeconomic background and achievement, New York, 1972, Seminar Press, Inc.

Dworkin, E. P., and Dworkin, A. L.: The activist counselor, Personn. Guid. J. **49:**748-753, 1971.

Egan, G.: Encounter: group processes for interpersonal growth, Belmont, Calif., 1970, Brooks/Cole Publishing Co.

Ellis, A.: The theory and practice of rational-emotive psychotherapy, New York, 1964, Lyle Stuart, Inc.

Ellis, A.: Reason and emotion in psychotherapy, New York, 1962, Lyle Stuart, Inc.

English, H. B., and English, A. C.: A comprehensive dictionary of psychological and psychoanalytical terms, New York, 1958, Longmans, Green & Co., Inc.

Ennis, B., and Siegel, L.: The rights of mental patients, New York, 1973, Avon Books.

Erikson, E. H.: Identity: youth and crisis, New York, 1968, W. W. Norton & Co., Inc.

Erikson, E. H.: Childhood and society, ed. 2, New York, 1963, W. W. Norton & Co., Inc. (ed. 1, 1950).

Erikson, E. H., editor: Youth: change and challenge, New York, 1963, Basic Books, Inc., Publishers.

Ethical practice: preserving human dignity, Personn. Guid. J. **50:**special issue, 1971.

Fairweather, G. W., et al.: Community life for the mentally ill: an alternative to institutional care, Chicago, 1969, Aldine Publishing Co.

Fanon, F.: The wretched of the earth, New York, 1968, Grove Press, Inc.

Fine, R.: Interpretation: the patient's response. In Hammer, E. F., editor: Use of interpretation in treatment, New York, 1968, Grune & Stratton, Inc.

Fisher, W., Mehr, J., and Truckenbrod, P.: Power, greed and stupidity in the mental health racket, Philadelphia, 1973, The Westminster Press.

Folsom, J. C.: Reality orientation for the elderly mental patient, J. Geriat. Psychiatry **1:**291-307, 1968.

Frankl, V. F.: Man's search for meaning, Boston, 1963, Beacon Press.

Fretz, B. R.: Postural movements in a counseling dyad, J. Counsel. Psychol. **13:**335-343, 1966.

Freud, S. A.: General introduction to psychoanalysis (translated by J. Riviere), New York, 1938, Garden City Publishing Co.

Fromm, E.: The art of loving, New York, 1956, Harper & Row, Publishers.

Fromm, E.: Escape from freedom, New York, 1941, Rinehart & Co.

Gardner, J.: Sexist counseling must go, Personn. Guid. J. **49:**705-714, 1971.

Gardner, M.: A design for continuing education with mental health associates, Baltimore, Md., Feb., 1973, Division of Manpower Development and Training, State Department of Health & Mental Hygiene. (Mimeographed.)

Gardner, R. A.: Therapeutic communication with children: the mutual storytelling technique, New York, 1971, Science House, Inc.

Ginott, H. G.: Between parent and teenager, New York, 1971, Avon Books.

Ginott, H. G.: Between parent and child, New York, 1965, The Macmillan Co.

Glasser, W.: Reality therapy: a new approach to psychiatry, New York, 1965, Harper & Row, Publishers.

Graham, R., and Valentine, M.: Alienation through isolation, Personn. Administr. **32**(2): 17-20, 1969.

Greenspoon, J.: The reinforcing effects of two spoken sounds on the frequency of two responses, Am. J. Psychol. **68:**409-416, 1955.

Grier, W. H., and Price, M. C.: Black rage, New York, 1968, Basic Books, Inc., Publishers.

Grinker, R. R., Sr., et al.: Psychiatric social work: a transactional case book, New York, 1961, Basic Books, Inc., Publishers.

Grosser, C., Henry, W. E., and Kelley, J. G., editors: Nonprofessionals in the human services, San Francisco, 1969, Jossey-Bass, Inc., Publishers.

Grossman, H. J., editor: Manual on terminology and classification in mental retardation, Washington, D. C., 1973, American Association on Mental Deficiency.

Guerney, B. J., Jr.: Psychotherapeutic agents: new roles for nonprofessionals, parents and teachers, New York, 1969, Holt, Rinehart & Winston, Inc.

Haase, R. F., and DiMattia, D. J.: Proxemic behavior: counselor, administrator and client preference for seating arrangement in dyadic interaction, J. Counsel. Psychol. **17:**319-325, 1970.

Haase, R. F., and Tepper, D. T., Jr.: Nonverbal components of emphatic communication, J. Counsel. Psychol. **19:**417-424, 1972.

Hadley, J. M., True, J. E., and Kepes, S. Y.:

An experiment in the education of the pre-professional mental health worker: the Purdue program, Indiana, 1968. (Unpublished manuscript.)

Hall, E. T.: The hidden dimension, New York, Garden City, N. Y., 1966, Doubleday & Co., Inc. (Paperback edition by Anchor Press.)

Hall, E. T.: The silent language, Garden City, N. Y., 1959, Doubleday & Co., Inc. (Paperback edition by Fawcett.)

Halleck, S.: The politics of therapy, New York, 1971, Science House, Inc.

Hammer, E. F., editor: Use of interpretation in treatment, New York, 1968, Grune & Stratton, Inc.

Harper, R. A.: Psychoanalysis and psychotherapy: 36 systems, Englewood Cliffs, N. J., 1959, Prentice-Hall, Inc.

Harris, T.: I'm OK—You're OK, New York, 1970, Harper & Row, Publishers.

Hasenfeld, Y., and English, R. A., editors: Human service organizations, Ann Arbor, 1973, The University of Michigan Press.

Hebert, M. R., and Schulman, M.: Educational change through violent dissent, Viewpoints: Bull. School Educ. Ind. Univ. **48:**139-149, 1972.

Heine, R. W.: The Negro patient in psychotherapy, J. Clin. Psychol. **6:**373-376, 1950.

Hinsie, L. E., and Campbell, R. J.: Psychiatric dictionary, ed. 4, New York, 1970, Oxford University Press, Inc.

Hobbs, N.: Sources of gain in counseling and psychotherapy, Am. Psychol. **17:**18-34, 1962.

Hollis, F.: Casework: a psychosocial therapy, ed. 2, New York, 1972, Random House, Inc.

Hopf, J.: Ethics in practice: one woman's solutions, Personn. Guid. J. **51:**48-52, 1972.

Horney, K.: Neurosis and human growth, New York, 1954, W. W. Norton & Co., Inc.

Hunt, W. A., editor: Human behavior and its control, Cambridge, Mass., 1971, Schenkman Publishing Co., Inc.

The institutional guide to DHEW policy on protection of human subjects, Washington, D. C., 1971, U. S. Department of Health, Education, & Welfare.

Ivey, A. E.: Microcounseling: the counselor as trainer, Personn. Guid. J. **51:** 311-316, 1973.

Ivey, A. E.: Microcounseling innovations in interviewing training, Springfield, Ill., 1971, Charles C Thomas, Publisher.

Janis, I. L., et al.: Personality: dynamics, development and assessment, New York, 1969, Harcourt, Brace & World, Inc.

Johnson, D. W.: Effects of the order of expressing warmth and anger on the actor and the listener, J. Counsel. Psychol. **18:**571-578, 1971a.

Johnson, D. W.: The effects of warmth of interaction, accuracy of understanding, and the proposal of compromises on the listener's behavior, J. Counsel. Psychol. **18:**207-216, 1971b.

Johnson, D. W., and Noonan, P. M.: Effects of acceptance and reciprocation of self-disclosures on the development of trust, J. Counsel. Psychol. **19:**411-416, 1972.

Johnson, W.: People in quandries: The semantics of personal adjustment, New York, 1946, Harper & Row, Publishers.

Joint Commission on Mental Health: challenge for the 1970's, New York, 1970, Harper & Row, Publishers.

Jorgensen, G. T., and Weigel, R. G.: Training psychotherapists: practices regarding ethics, personal growth and focus of responsibility, Profess. Psychol. **4:**23-27, 1973.

Joseph, S. M.: The me nobody knows, New York, 1969, Avon Books.

Jourard, S.: The transparent self, New York, 1971, D. Van Nostrand Co.

Jourard, S. M., and Jaffe, P. E.: Influence of an interviewer's disclosure on the self-disclosing behavior of interviewees, J. Counsel. Psychol. **17:**252-259, 1970.

Kadushin, A.: The social work interview, New York, 1972, Columbia University Press.

Kahn, R. L., and Cannell, C. F.: The dynamics of interviewing, New York, 1957, John Wiley & Sons, Inc.

Kaslow, F. W., et al.: Issues in human services, San Francisco, 1972, Jossey-Bass, Inc., Publishers.

Kaul, T. J., and Schmidt, L. D.: Dimensions of interviewer trustworthiness, J. Counsel. Psychol. **18:**542-548, 1971.

Kaul, T. J., Kaul, M. J., and Bednar, R. L.: Counselor confrontation and clinical depth of self-exploration, J. Counsel. Psychol. **20:**132-136, 1973.

Keeley, S. M., Shemberg, K. M., and Ferber, H.: The training and use of undergraduates as behavior analysts in the consultative process, Profess. Psychol. **4:**59-63, 1973.

Kell, B., and Mueller, W.: Impact and change: a study of counseling relationships, New York, 1966, Appleton-Century-Crofts.

King, J. B.: Community college mental health worker in the South, Atlanta, 1970, Southern Regional Education Board.

Konopka, G.: Effective communication with adolescents in institutions, New York, 1965, Child Welfare League of America.

Kovar, L. C.: Faces of the adolescent girl, Englewood Cliffs, N. J., 1968, Prentice-Hall, Inc.

Krumboltz, J. D., and Thoresen, C. E.: Behavioral counseling, New York, 1969, Holt, Rinehart & Winston, Inc.

Kuriloff, P. J.: The counselor as psychoecologist, Personn. Guid. J. **51:**321-327, 1973.

Laing, R. D.: The politics of experience, New York, 1967, Pantheon Books, Inc.

Laing, R.: The divided self, London, 1960, Tavistock Publications, Ltd.

Lesser, W. M.: The relationship between counseling progress and empathic understanding, J. Counsel. Psychol. **8:**330-336, 1961.

Levy, R. B.: Self-revelation through relationships, Englewood Cliffs, N. J., 1972, Prentice-Hall, Inc.

Lewis, H. R., and Streitfeld, H. S.: Growth games: how to tune into yourself, your family, your friends, New York, 1971, Bantam Books, Inc.

Lieb, J. Lipsitch, I., and Slaby, A.: The crisis team, New York, 1973, Harper & Row, Publishers.

Loughary, J. Friesen, D., and Hurst, R.: Autocom: a computer-based automated counseling simulation system, Personn. Guid. J. **45:**6-15, 1966.

Luft, J.: Of human interaction, Palo Alto, Calif., 1969, National Press Books.

Lukas, J. A.: Don't shoot, we are your children, New York, 1972, Dell Publishing Co., Inc.

MacKinnon, R. A., and Michels, R.: The psychiatric interview in clinical practice, Philadelphia, 1971, W. B. Saunders Co.

Magoon, T. M., Golann, S. E., and Freeman, R. W.: Mental health counselors at work, New York, 1969, Pergamon Press, Inc.

Malamud, D. I., and Machover, S.: Toward self-understanding: group techniques in self-confrontation, Springfield, Ill., 1965, Charles C Thomas, Publisher.

Mallott, R. W.: Contingency management, Kalamazoo, Mich., 1972, Behaviordelia.

Markeny, K.: Counselors as environmental engineers, Personn. Guid. J. **49:**439-444, 1971.

Maslow, A. H.: The farther reaches of human nature, New York, 1971, The Viking Press, Inc.

Maslow, A. H.: Motivation and personality, ed. 2, New York, 1970, Harper & Row, Publishers (ed. 1, 1954).

Maslow, A. H.: Toward a psychology of being, ed. 2, Princeton, N. J., 1968, Van Nostrand Reinhold Co.

Matarazzo, J. D.: Changing concepts: care and caregivers, three points of view, Ment. Hyg. **52:**163-164, 1968.

Matarazzo, J. D., Saslow, G., and Matarazzo, R.: The interaction chronograph as an instrument for objective measurement of interaction patterns during interviews, J. Psychol. **41:**347-367, 1965.

May, E. P.: Quantity or quality in dealing with human problems, Personn. Guid. J. **49:**376-382, 1971.

May, R.: Love and will, New York, 1969, W. W. Norton & Co., Inc.

McLuhan, M., and Fiore, Q.: The medium is the message, New York, 1967, Bantam Books, Inc.

McMullin, R. E.: Effects of counselor focusing on client self-experiencing under low attitudinal conditions, J. Counsel. Psychol. **19:**282-285, 1972.

McPheeters, H. L., and Baker, E. J.: Community college programs in mental health technology, Atlanta, 1969, Southern Regional Education Board.

McPheeters, H. L., and King, J. B.: Plans for teaching mental health workers, Atlanta, 1971, Southern Regional Education Board.

Mehrabian, A.: Significance of posture and position in the communication of attitude and status relationships, Psychol. Bull. **71:**359-372, 1969.

Mehrabian, A.: Communication without words, Psychol. Today **2**(4):53-55, 1968.

Mickelson, D. J., and Stevic, R. R.: Differential effects of facilitative and nonfacilitative behavioral counselors, J. Counsel. Psychol. **18:**314-319, 1971.

Middleman, R. R.: The non-verbal method in working with groups, New York, 1968, Association Press.

Millet, K.: Sexual politics, Garden City, N. Y., 1970, Doubleday & Co., Inc.

Milne, A. A.: When we were very young, New York, 1924, E. P. Dutton & Co., Inc.

Morgan, R., editor: Sisterhood is powerful, New York, 1970, Random House, Inc.

Morris, D.: Intimate behavior, New York, 1971, Random House, Inc.

Mozee, E.: Counselor, evaluate thyself, Personn. Guid. J. **51:**285-287, 1972.

Mullen, J., and Abeles, N.: Relationship of liking, empathy, and therapist's experience to outcome of therapy, J. Counsel. Psychol. **19:**121-124, 1972.

Murphy, K. C., and Strong, S. R.: Some effects of simulating self-disclosure, J. Counsel. Psychol. **19:**121-124, 1972.

Myerson, P. G.: The meanings of confrontation. In Adler, G., and Myerson, P. G.: Confrontation in psychotherapy, New York, 1973, Science House, Inc.

Olesker, W., and Balter, L.: Sex and empathy, J. Consult. Psychol. **19:**559-562, 1972.

Parker, T.: The twisting lane: the hidden world of sex offenders, New York, 1972, Harper & Row, Publishers.

Patterson, D. H.: Theories of counseling and psychotherapy, rev. ed., New York, 1973, Harper & Row, Publishers.

Payne, P. A., Weiss, S. D., and Kapp, R. A.: Didactic, experiential, and modeling factors in the learning of empathy, J. Counsel. Psychol. **19:**425-429, 1972.

Perls, F., Hefferline, R., and Goodman, P.: Gestalt therapy, New York, 1951, Julian Press, Inc.

Pervin, L. A.: Personality: theory, assessment, and research, New York, 1970, John Wiley & Sons, Inc.

Peterfreund, E.: Information systems and psychoanalysis, New York, 1971, International Universities Press.

Phillips, E. L., and Wiener, D. N.: Short-term psychotherapy and structured behavior change, New York, 1966, McGraw-Hill Book Co.

Piaget, J.: The origins of intelligence in children (translated by M. Cook), ed. 2, New York, 1952, International Universities Press.

Prescott, D.: The child in the educative process, New York, 1957, McGraw-Hill Book Co.

Previn, D.: On my way to where, New York, 1971, Bantam Books, Inc.

Price, R., and Denner, B.: The making of a mental patient, New York, 1973, Holt, Rinehart & Winston, Inc.

Rado, S.: Adaptational psychodynamics: motivation and control, New York, 1969, Science House, Inc.

Ream, C.: Youth culture: humanity's last chance, Personn. Guid. J. **49:**699-704, 1971.

Reik, T.: Listening with the third ear, New York, 1949, Farrar, Straus & Giroux, Inc.

Ribble, M.: The personality of the young child: an introduction for puzzled parents, New York, 1955, Columbia University Press.

Riese, H.: Heal the hurt child: an approach through educational therapy with special reference to the extremely deprived Negro child, Chicago, 1962, University of Chicago Press.

Riessman, F.: Are the deprived non-verbal? In Riessman, F., Cohen, J., and Pearl, A., editors: Mental health of the poor: new treatment approaches for low income people, New York, 1964, The Free Press.

Riessman, F., Cohen, J., and Pearl, A., editors: Mental health of the poor: new treatment approaches for low income people, New York, 1964, The Free Press.

Rimland, B.: Infantile autism, New York, 1964, Appleton-Century-Crofts.

Rioch, M. J., Elkes, C., and Flint, A.: Pilot project in training mental health counselors, Chevy Chase, Md., 1965, National Institute of Mental Health.

Rogers, C. R.: Encounter groups, New York, 1970, Harper & Row, Publishers.

Rogers, C. R.: On becoming a person: a therapist's view of psychotherapy, Boston, 1961, Houghton Mifflin Co.

Rogers, C. R.: The characteristics of a helping relationship, Personn. Guid. J. **37:**6-16, 1958.

Rogers, C. R.: Client-centered therapy, Boston, 1951, Houghton Mifflin Co.

Roles and functions for different levels of mental health workers, Atlanta, 1969, Southern Regional Education Board.

Roll, W. V., Schmidt, L. D., and Kaul, T. J.: Perceived interviewer trustworthiness among black and white convicts, J. Counsel. Psychol. **19:**537-541, 1972.

Rosenbaum, M., and Berger, M., editors: Group psychotherapy and group function, New York, 1963, Basic Books, Inc., Publishers.

Roszak, T.: The making of a counter culture, Garden City, N. Y., 1969, Doubleday & Co., Inc.

Ruesch, J.: Therapeutic communication, New York, 1965, W. W. Norton & Co., Inc.

Sager, C. J., Brayboy, T. L., and Waxenberg, B. R.: Black ghetto family in therapy, New York, 1970, Grove Press, Inc.

Satir, V.: Peoplemaking, Palo Alto, Calif., 1972, Science & Behavior Books, Inc.

Satir, V.: Cojoint family therapy: a guide to theory and technique, Palo Alto, Calif., 1964, Science & Behavior Books. Inc.

Saulnier, L., and Simard, T.: Personal growth and interpersonal relations, Englewood Cliffs, N. J., 1973, Prentice-Hall, Inc.

Scheflen, A.: Communicational structure: analysis of a psychotherapy transaction, Bloomington, Ind., 1973, Indiana University Press.

Schmidt, L. D., and Strong, A. R.: Attractiveness and influence in counseling, J. Counsel. Psychol. **18:**348-351, 1971.

Schulberg, H. C.: Challenge of human service programs for psychologists, Am. Psychol. **27:**566-573, 1972.

Schulberg, H. C., Baker, F., and Roen, S. R., editors: Developments in human services, New York, 1973, Behavioral Publications, Inc., vol. 1.

Schulman, E. D.: From a social interaction focus to a mental health technician worker, Presented at the American Psychological Association meeting, 1968.

Schulman, E. D.: Mental health technician: the new professional, Presented at the Maryland Psychological Association meeting, 1967.

Schulman, E. D.: Baltimore's social interaction program, Jr. College J. **37:**34-36, Dec., 1966.

Schumacher, L. E., Banikiotes, P. G., and Banikiotes, F. G.: Language compatibility and minority group counseling, J. Counsel. Psychol. **19:**255-256, 1972.

Schutz, W. C.: Joy, New York, 1967, Grove Press, Inc.

Sells, S. B., editor: The definition and measurement of mental health, Washington, D. C., 1968, U. S. Department of Health, Education, & Welfare.

Selye, H.: The stress of life, New York, 1956, McGraw-Hill Book Co.

Shapiro, J. D., Foster, C. P., and Powell, T.: Facial and bodily cues of genuineness, empathy and warmth, J. Clin. Psychol. **24**:233-236, 1968.

Shimkunas, A. M.: Demand for intimate self-disclosure and pathological verbalizations in schizophrenia, J. Abnorm. Psychol. **80**:197-205, 1972.

Shneidman, E. S., editor: Essays in self-destruction, New York, 1967, Jason Aronson, Inc.

Shostrom, E. L.: Man the manipulator, Nashville, Tenn., 1967, Abingdon Press.

Shostrom, E. L.: Personal orientation inventory, San Diego, 1963, Educational & Industrial Testing Service.

Simmons, J. E.: Psychiatric examination of children, Philadelphia, 1969, Lea & Febiger.

Simon, R.: The paraprofessionals are coming, Presented at the American Psychological Association meeting, 1970.

Skinner, B.: Beyond freedom and dignity, New York, 1971, Bantam Books, Inc.

Smith, P. M., Jr.: Black activists for liberation, not guidance, Personn. Guid. J. **49**:721-726, 1971.

Smith, W. D., and Martinson, W. D.: Counselors' and counselees' learning style on interview behavior, J. Counsel. Psychol. **18**:138-141, 1971.

Sobey, F.: The nonprofessional revolution in mental health, New York, 1970, Columbia University Press.

Social change and the mental health of children. Report of Task Force VI and excerpts from the report of the Committee on Children of Minority Groups, Joint Commission on Mental Health of Children, New York, 1973, Harper & Row, Publishers.

Sommer, R.: Personal space: the behavioral basis of design, Englewood Cliffs, N. J., 1969, Prentice-Hall, Inc.

Sperry, L.: Counselors and learning styles, Personn. Guid. J. **51**:478-483, 1973.

Spitz, R. A.: The first year of life, New York, 1965, International Universities Press.

STASH notes: Phencyclidine (PCP), STASH Capsules **5**(2):1-2, 1973.

Stillman, S., and Resnick, H.: Does counselor attire matter? J. Counsel. Psychol. **19**:347-348, 1972.

Strong, S. R., and Dixon, D. N.: Expertness, attractiveness, and influence in counseling, J. Counsel. Psychol. **18**:562-570, 1971.

Strong, S. R., et al.: Nonverbal behavior and perceived counselor characteristics, J. Counsel. Psychol. **18**:554-561, 1971.

Sullivan, H. S.: The interpersonal theory of psychiatry (edited by H. S. Perry and M. L. Gowel), New York, 1953, W. W. Norton & Co., Inc.

Szasz, T.: Ideology and insanity: essays on the psychiatric dehumanization of man, Garden City, N. Y. 1970, Doubleday & Co., Inc.

Szasz, T. S.: The myth of mental illness, New York, 1961, Harper & Row, Publishers.

Talkington, P. C.: Critical issues in psychiatry: a call for a reassessment of our nation's mental health care, Hosp. Community Psychiatry **24**:17-22, 1973.

Taylor, J. L., and Walford, R.: Simulation in the classroom, Harmondsworth, England, 1972, Penquin Publishing Co., Ltd.

Tharp, R. G., and Wetzel, R. J.: Behavior modification in the natural environment, New York, 1969, Academic Press, Inc.

Thompson, T., and Grabowski, J., editors: Behavior modification of the mentally retarded, New York, 1972, Oxford University Press, Inc.

Toffler, A.: Future shock, New York, 1970, Random House, Inc.

Truax, C. B.: Some implications of behavior therapy for psychotherapy, J. Counsel. Psychol. **13**:160-170, 1966.

Truax, C. B., and Carkhuff, R. R.: Toward effective counseling and psychotherapy, Chicago, 1967, Aldine Publishing Co.

Truax, C. B., and Carkhuff, R. R.: An introduction to counseling and psychotherapy: training and practice, Chicago, 1966, Aldine Publishing Co.

True, J., Young, C., and Packard, M.: A national survey of associate degree programs in mental health, Columbia, Md., 1972, The Center for Human Services, The Johns Hopkins University. (Unpublished manuscript.)

Twerski, A.: Alcologia, Behavior Today **4**(19):2, 1973.

Ullman, L., and Krasner, L., editors: Case studies in behavior modification, New York, 1966, Holt, Rinehart & Winston, Inc.

Van Riper, B. W.: Toward a separate professional identity, J. Counsel. Psychol. **19**:117-120, 1972.

Verville, E.: Behavior problems of children, Philadelphia, 1967, W. B. Saunders Co.

Vidaver, R. M.: The mental health technician: Maryland's design for a new health career, Am. J. Psychiatry **125**:1013-1023, 1969.

Wagner, N., and Haug, M. J.: Chicanos, St. Louis, 1971, The C. V. Mosby Co.

Washington Post: Gallup Poll, Jan. 30, 1973.

Webster, S. D.: Humanness: the one essential, Personn. Guid. J. **51**:378-379, 1973.

Wellner, A. M., and Simon, R.: A survey of

associate-degree programs for mental health technicians, Washington, D. C., 1969, U. S. Department of Health, Education, & Welfare.

White, O. R.: A glossary of behavioral terminology, Champaign, Ill., 1971, Research Press Co.

Widgery, R., and Stackpole, C.: Desk positions, interviewee anxiety and interviewer credibility: an example of cognitive balance in a dyad, J. Counsel. Psychol. **19**:173-177, 1972.

Wiener, M., et al.: Nonverbal behavior and nonverbal communication, Psychol. Rev. **79**:185-214, 1972.

Wolpe, J.: The practice of behavior therapy, New York, 1969, Pergamon Press, Inc.

Wolpe, J., and Lazurus, A. A.: Behavior therapy techniques, New York, 1966, Pergamon Press, Inc.

Woody, R. H.: Psychobehavioral counseling and therapy: integrating behavioral and insight therapy, New York, 1971, Appleton-Century-Crofts.

Wortis, J., editor: Mental retardation and developmental disabilities, New York, 1973, Brunner/Mazel, Inc., vol. 5.

Wrenn, C.: The counselor in the changing world, Washington, D. C., 1962, American Personnel & Guidance Association.

Wurster, C. R., editor: Statistics in mental health, Washington, D. C., 1970, U. S. Department of Health, Education, & Welfare.

Wyss, D.: Psychoanalytic schools from the beginning to the present, (translated by G. Onn), New York, 1973, Jacob Aronson, Inc.

Yamamoto, J., et al.: Racial factors in patient selection, Am. J. Psychiatry **124**:630-636, 1967.

Zimmer, J., and Anderson, S.: Dimensions of positive regard and empathy, J. Counsel. Psychol. **15**:417-426, 1968.

Index

DATE DUE

NOV 1 2 '81	NOV 12 '81		
DEC 3 '81	DEC 7 '81		
DEC 1 0 '81	DEC 10 '81		
GAYLORD			PRINTED IN U.S.A